Protein and Pharmaceutical Engineering

UCLA Symposia on Molecular and Cellular Biology, New Series

Series Editor, C. Fred Fox

RECENT TITLES

Volume 108
Acute Lymphoblastic Leukemia, Robert Peter Gale and Dieter Hoelzer, *Editors*

Volume 109
Frontiers of NMR in Molecular Biology, David Live, Ian M. Armitage, and Dinshaw J. Patel, *Editors*

Volume 110
Protein and Pharmaceutical Engineering, Charles S. Craik, Robert Fletterick, C. Robert Matthews, and James Wells, *Editors*

Volume 111
Glycobiology, Ernest G. Jaworski and Joseph R. Welply, *Editors*

Volume 112
New Directions in Biological Control: Alternatives for Suppressing Agricultural Pests and Diseases, Ralph R. Baker and Peter E. Dunn, *Editors*

Volume 113
Immunogenicity, Charles A. Janeway, Jr., Jonathan Sprent, and Eli Sercarz, *Editors*

Volume 114
Genetic Mechanisms in Carcinogenesis and Tumor Progression, Curtis Harris and Lance A. Liotta, *Editors*

Volume 115
Growth Regulation of Cancer II, Marc E. Lippman and Robert B. Dickson, *Editors*

Volume 116
Transgenic Models in Medicine and Agriculture, Robert B. Church, *Editor*

Volume 117
Early Embryo Development and Paracrine Relationships, Susan Heyner and Lynn M. Wiley, *Editors*

Volume 118
Cellular and Molecular Biology of Normal and Abnormal Erythroid Membranes, Carl M. Cohen and Jiri Palek, *Editors*

Volume 119
Human Retroviruses, Jerome E. Groopman, Irvin S.Y. Chen, Myron Essex, and Robin A. Weiss, *Editors*

Volume 120
Hematopoiesis, David W. Golde and Steven C. Clark, *Editors*

Volume 121
Defense Molecules, John J. Marchalonis and Carol L. Reinisch, *Editors*

Volume 122
Molecular Evolution, Michael T. Clegg and Stephen J. O'Brien, *Editors*

Volume 123
Molecular Biology of Aging, Caleb E. Finch and Thomas E. Johnson, *Editors*

Volume 124
Papillomaviruses, Peter M. Howley and Thomas R. Broker, *Editors*

Volume 125
Developmental Biology, Eric H. Davidson, Joan V. Ruderman, and James W. Posakony, *Editors*

Volume 126
Biotechnology and Human Genetic Predisposition to Disease, Charles R. Cantor, C. Thomas Caskey, Leroy E. Hood, Daphne Kamely, and Gilbert S. Omenn, *Editors*

Volume 127
Molecular Mechanisms in DNA Replication and Recombination, Charles C. Richardson and I. Robert Lehman, *Editors*

Volume 128
Nucleic Acid Methylation, Gary A. Clawson, Dawn B. Willis, Arthur Weissbach, and Peter A. Jones, *Editors*

Volume 129
Plant Gene Transfer, Christopher J. Lamb and Roger N. Beachy, *Editors*

Volume 130
Parasites: Molecular Biology, Drug and Vaccine Design, Nina M. Agabian and Anthony Cerami, *Editors*

Volume 131
Molecular Biology of the Cardiovascular System, Robert Roberts and Joseph F. Sambrook, *Editors*

Volume 132
Obesity: Towards a Molecular Approach, George A. Bray, Daniel Ricquier, and Bruce M. Spiegelman, *Editors*

Volume 133
Structural and Organizational Aspects of Metabolic Regulation, Paul A. Srere, Mary Ellen Jones, and Christopher K. Mathews, *Editors*

Please contact the publisher for information about previous titles in this series.

Protein and Pharmaceutical Engineering

Proceedings of an Upjohn-UCLA Symposium, Held at Park City, Utah, January 17-22, 1989

Editors

Charles S. Craik
Departments of Pharmaceutical Chemistry, Biochemistry and Biophysics
University of California-San Francisco
San Francisco, California

Robert Fletterick
Department of Biochemistry
University of California-San Francisco
San Francisco, California

C. Robert Matthews
Pennsylvania State University
University Park, Pennsylvania

James Wells
Department of Biomolecular Chemistry
Genentech
South San Francisco, California

WILEY-LISS

A JOHN WILEY & SONS, INC., PUBLICATION
New York • Chichester • Brisbane • Toronto • Singapore

Address all Inquiries to the Publisher
Wiley-Liss, Inc., 41 East 11th Street, New York, NY 10003

Printed in United States of America

The publication of this volume was facilitated by the authors and editors who submitted the text in a form suitable for direct reproduction without subsequent editing or proofreading by the publisher.

Library of Congress Cataloging-in-Publication Data

Upjohn-UCLA Symposium on Protein and Pharmaceutical Engineering (1989 : Park City, Utah)
Protein and pharmaceutical engineering : proceedings of a UCLA Symposium held at Park City, Utah January 17-22, 1989 / editors, Charles S. Craik ... et al.].
p. cm. -- (UCLA symposia on molecular and cellular biology ; new ser., v. 110)
"Upjohn-UCLA Symposium on Protein and Pharmaceutical Engineering"--Pref.
Includes bibliographical references.
ISBN 0-471-56771-X
1. Protein engineering--Congresses. 2. Pharmaceutical chemistry--Congresses. I. Craik, Charles S. II. Upjohn Company. III. University of California, Los Angeles. IV. Title. V. Series.
[DNLM: 1. Chemistry, Pharmaceutical--congresses. 2. Proteins--congresses. W3 U17N v. 110 / QU 55 U67 1989p]
TP248.65.P76U65 1989
660'.63--dc20
DNLM/DLC
for Library of Congress 90-11911
CIP

Contents

Contributors

Mark D. Adams, Department of Biological Chemistry, University of Michigan, Ann Arbor, 48109 **[145]**

Lilia M. Babé, Departments of Pharmaceutical Chemistry and Biochemistry and Biophysics, University of California at San Francisco, San Francisco, CA 94143 **[71]**

Philip J. Barr, Chiron Corporation, Emeryville, CA 94608 **[71]**

Ian C. Bathurst, Chiron Corporation, Emeryville, CA 94608 **[71]**

Patricia A. Benkovic, Department of Chemistry, The Pennsylvania State University, University Park, PA 16802 **[89]**

Stephen J. Benkovic, Department of Chemistry, The Pennsylvania State University, University Park, PA 16802 **[89]**

Stephen C. Blacklow, Department of Chemistry, Harvard University, Cambridge, MA 02138 **[159]**

Philip H. Bolton, Department of Chemistry, Wesleyan University, Middletown, CT 06457 **[17]**

Ronald Breslow, Department of Chemistry, Columbia University, New York, NY 10027 **[135]**

Janet C. Cheetham, Laboratory of Molecular Biophysics, University of Oxford, Oxford OX1 3QU, England **[35]**

Charles S. Craik, Departments of Pharmaceutical Chemistry, Biochemistry and Biophysics, University of California at San Francisco, San Francisco, CA 94143 **[xiii, 71]**

Mark Dell'Acqua, Department of Chemistry and Biochemistry, University of Maryland, College Park, MD 20742 **[17]**

Edward Eisenstein, Departments of Molecular and Cell Biology and Biochemistry, Virus Laboratory, University of California at Berkeley, Berkeley, CA 94720; present address: Center for Advanced Research in Biotechnology University of Maryland, Rockville, MD 20850 **[167]**

Robert Fletterick, Department of Biochemistry, University of California at San Francisco, San Francisco, CA 94143 **[xiii]**

The numbers in brackets are the opening page numbers of the contributors' articles.

Robert O. Fox, Department of Molecular Biophysics and Biochemistry, The Howard Hughes Medical Institute, Yale University, New Haven, CT 06511 [1]

John A. Gerlt, Department of Chemistry and Biochemistry, University of Maryland, College Park, MD 20742 [17]

John F. Gill, Department of Molecular Biophysics and Biochemistry, Yale University, New Haven, CT 066511 [1]

Jonathan M. Goldberg, Department of Biochemistry, University of California at Berkeley, Berkeley, CA 94720 [105]

David W. Hilber, Department of Chemistry and Biochemistry, University of Maryland, College Park, MD 20742 [17]

Roger A. Kautz, Department of Molecular Biophysics and Biochemistry, Yale University, New Haven, CT 06511 [1]

Jack F. Kirsch, Department of Biochemistry, University of California at Berkeley, Berkeley, CA 94720 [105]

Jeremy R. Knowles, Department of Chemistry, Harvard University, Cambridge, MA 02138 [159]

R. Luise Krauth-Siegel, Institut fur Biochemie II, Medizinische Fakultat der Universitat, D-6900 Heidelberg, Federal Republic of Germany [119]

Eaton Lattman, Department of Biophysics, Johns Hopkins University School of Medicine, Baltimore, MD 21205 [17]

Richard A. Lerner, Department of Molecular Biology, Research Institute of Scripps Clinic, La Jolla, CA 92037 [89]

Patrick Loll, Department of Biophysics, Johns Hopkins University School of Medicine, Baltimore, MD 21205 [17]

David J. Maguire, Department of Biological Chemistry, University of Michigan, Ann Arbor, MI 48109; present address: Division of Science and Technology, Griffith University, Brisbane Q4111, Australia [145]

David W. Markby, Departments of Molecular and Cell Biology and Biochemistry, Virus Laboratory, University of California at Berkeley, Berkeley, CA 94720 [167]

Andrew C.R. Martin, Laboratory of Molecular Biophysics, University of Oxford, Oxford OX1 3QU, England [35]

Frank R. Masiarz, Chiron Corporation, Emeryville, CA 94608 [71]

C. Robert Matthews, Pennsylvania State University, University Park, PA 16802 [xiii]

Alok K. Mitra, Department of Biochemistry and Biophysics, University of California at San Francisco, San Francisco, CA 94143 [55]

Andrew D. Napper, Department of Chemistry, The Pennsylvania State University, University Park, PA 16802 [89]

Dale L. Oxender, Department of Biological Chemistry, University of Michigan, Ann Arbor, MI 48109 [145]

Emil F. Pai, Abteilung Biophysik, Max-Planck-Institut fur Medizinische Forschung, D-6900 Heidelberg, Federal Republic of Germany [119]

Sergio Pichuantes, Departments of Pharmaceutical Chemistry and Biochemistry and Biophysics, University of California at San Francisco, San Francisco, CA 94143 **[71]**

David L. Pompliano, Department of Chemistry, Harvard University, Cambridge, MA 02138 **[159]**

Tayebeh Pourmotabbed, Department of Chemistry and Biochemistry, University of Maryland, College Park, MD 20742 **[17]**

Anthony R. Rees, Laboratory of Molecular Biophysics, University of Oxford, Oxford OX1 3QU, England **[35]**

Sally Roberts, Laboratory of Molecular Biophysics, University of Oxford, Oxford OX1 3QU, England **[35]**

Howard K. Schachman, Departments of Molecular and Cell Biology and Biochemistry, Virus Laboratory, University of California at Berkeley, Berkeley, CA 94720 **[167]**

Michael J. Shuster, Department of Biochemistry and Biophysics, University of California at San Francisco, San Francisco, CA 94143 **[55]**

Susan M. Stanczyk, Department of Chemistry, Wesleyan University, Middletown, CT 06457 **[17]**

Robert M. Stroud, Department of Biochemistry and Biophysics, University of California at San Francisco, San Francisco, CA 94143 **[55]**

Francis X. Sullivan, Department of Biological Chemistry and Molecular Pharmacology, Harvard Medical School, Boston, MA 02115 **[119]**

Michael D. Toney, Department of Biochemistry, University of California at Berkeley, Berkeley, CA 94720 **[105]**

Christopher T. Walsh, Department of Biological Chemistry and Molecular Pharmacology, Harvard Medical School, Boston, MA 02115 **[119]**

James Wells, Department of Biomolecular Chemistry, Genentech, South San Francisco, CA 94080 **[xiii]**

Preface

Protein engineering is beginning to live up to its name. The 1989 Upjohn–UCLA Symposium on **Protein and Pharmaceutical Engineering** held in late January 1989, in the magnificent winter setting of the Wasatch Mountains, Park City, Utah, attested to this development. It was only five years ago when the term "protein engineering" was coined and found its way into the literature. The term was denounced by investigators in the field as a misnomer, because it implied that a protein engineer knew what he or she was doing. The practice of protein engineering was viewed as akin to protein terrorism, since frequently it wreaked havoc on unsuspecting proteins; it was also considered a very expensive way to inactivate proteins. Perhaps the most telling denunciation was the analogy between a modern chemist working with a protein and a small boy in a cage containing a wild animal—to call the chemist a protein engineer was like calling the boy an animal trainer. However, recent advances, many of which were presented at the meeting and are documented in this volume, suggest that the vision of the protein engineers have begun to sharpen, and have come to focus on understanding the proteins that are being manipulated. With this knowledge to draw from, real progress can now be made in designing proteins and engineer them for the development of pharmaceuticals.

The meeting brought together a broad cross-section of experts who shared the common goal of dissecting the architecture and activity of a protein to uncover underlying principles of form and function.

"Focus on first principles" was a common theme in many of the presentations. For without that type of understanding, the ultimate goals of *de novo* design or rational redesign will be out of reach, and the burgeoning field that holds such great promise will remain mere wishful thinking. The meeting was attended by over 350 participants; it was organized into morning and late afternoon sessions, and the evenings assigned to poster sessions. There were two notable deviations from this schedule. The keynote address by Professor Jeremy Knowles entitled "Whither Protein Engineering?" was presented on the first evening of the meeting and was also attended by the participants of a concurrently held meeting on NMR Spectroscopy Macromolecules. The banquet address by Professor Howard Schachman entitled "Having Fun with Aspartate Transcarbamylase" was also an evening presentation. Both accounts of marvelous scientific sagas served to buttress the theme of the meeting, and provided wise perspectives for the participants. Elements of these addresses are represented in this volume.

The wide breadth of the meeting content was due to the diverse but common interests of the participants. By joining the forces of molecular genetics, enzymology, molecular modelling, NMR spectroscopy, and x-ray crystallography, substantial advances were made in understanding the roles played by specific amino acids in protein structure and function. Selected examples of the presentations are presented in this volume and fall into one of two broad categories: Physical Analysis of Protein Structure and Chemical Aspects of Protein Function. In the first category, x-ray crystallography and NMR spectroscopy are used to explore relationships between structure and function by analyzing the architecture of native and variant proteins. In the second category, chemical approaches, ranging from organic and biological synthesis to site-directed mutagenesis, are used to probe the function of proteins and enzymes. As our individual efforts merge to form a new discipline, volumes such as this one serve to mark our progress.

We wish to thank the Upjohn Company for its generous sponsorship of this meeting. We gratefully acknowledge additional support from Hoffmann-La Roche, Inc. We also thank the many individuals who made the meeting a success, among them Drs. Gary Neil and Ralph Christophersen of The Upjohn Company; the projectionists, Luke Evnin and John Vasquez; and the UCLA Symposia staff, Betty Handy, Robin Yeaton-Woo, Jackie Wester, and Hank Harwood.

Charles S. Craik
Robert Fletterick
C. Robert Matthews
James Wells

Protein and Pharmaceutical Engineering, pages 1–15

ASSIGNMENT OF HISTIDINE RESONANCES IN THE NMR SPECTRUM OF STAPHYLOCOCCAL NUCLEASE USING SITE-DIRECTED MUTAGENESIS

Roger A. Kautz[1], John F. Gill[1], Robert O. Fox[1,2]*

Department of Molecular Biophysics and Biochemistry[1], and
The Howard Hughes Medical Institute[2],
Yale University 260 Whitney Ave., P.O. Box 6666 New Haven, CT 06511

ABSTRACT Staphylococcal nuclease occurs in at least two folded conformations which are in slow exchange on the NMR time-scale. The kinetics have been demonstrated previously by magnetization transfer (MT) experiments using histidine C2 proton resonances. The assignment of the histidine C2 proton resonances is necessary for a structural interperetation of these results. We report the unambiguous assignment of the four histidine resonances by comparison of wild type protein to histidine substitution mutants over a range of pH.

INTRODUCTION

Staphylococcal nuclease (deoxyribonucleate 3' nucleotidohydrolase, EC 3.1.4.4) is a small globular protein of 149 residues with no disulfide bonds, cysteines or prosthetic groups which requires Ca^{2+} to hydrolyze DNA and RNA into 3'-phosphonucleotides and dinucleotides. The 3',5'-diphosphonucleotides (pdNp) bind to nuclease with high affinity and act as competitive inhibitors. The crystal structures of the apo-enzyme and the ternary complex with pdTp and Ca^{2+} have been reported (1,2,3) and have been further refined more recently (4,5). The histidine region of the NMR spectrum of unliganded nuclease shows evidence of a minor folded form of the protein in slow exchange with the major folded form at rates measurable by magnetization transfer techniques (6).

A detailed magnetization transfer analysis of the histidine resonances under conditions where the protein

This work was supported by NIH grant AI 23923

sample is near the middle of the thermal unfolding transition revealed two unfolded states in slow exchange on the NMR timescale (7). A hypothesis was advanced that the observed conformational heterogeneity involves a mixture of *cis* and *trans* isomers at the lys-116 pro-117 peptide (6). This hypothesis has been supported by analysis of a site directed mutant where the pro-117 residue has been replaced by a glycine resulting in a loss of the observed conformational heterogeneity (8). The kinetic measurements of the folding, unfolding and interconversion rates are also consistent with the *cis/trans* model. The four histidine C2 protons differ in the behavior of their major and minor resonances. We have assigned the NMR resonances to the chemical sequence and thus the three-dimensional structure to further map the site of the structural heterogeneity.

Resonance assignments have always been a formidable first problem in the study of macromolecules by NMR. Recently great progress has been made in the sequential assignment of high resolution NMR spectra of small globular proteins using 2D NMR methods (9,10). Histidine C2 protons can still pose a special assignment problem as they are not J-coupled to nonexchangeable protons and are sometimes not sufficiently close to other hydrogen atoms to display observable NOE effects. These methods are inefficient when only a few assignments are required (in this case the histidine C2 protons) since the confidence in any single assignment is high only when a large fraction of the assignments are made. Nuclease (149 residues) is larger than the largest proteins for which complete 2D sequential assignments have been reported: *E. coli* thioredoxin (11) (108 residues) and lysozyme (12) (129 residues).

The lack of general methods for selectively labelling individual histidine residues in a protein has necessitated a new approach for each new assignment based on unique features of each protein system. Histidine 12 in bovine pancreatic ribonuclease A (RNase A) was assigned by exploiting the ribonuclease S (RNase S) system. The single histidine (His-12) of the S-peptide was separately ^{2}H exchanged and recombined with unexchanged S protein (13). The spectrum of the reconstituted RNase S had lost resonance H2, identifying it as His-12. Ohe et al.(14) demonstrated a more general assignment technique for ribonuclease A. They measured the exchange rates of individual histidine C2 protons in ribonuclease A using partial ^{3}H exchange, following the incorporation of radioactivity in individual histidines by two dimensional gel analysis of characterized proteolytic fragments. The histidine C2 proton and exchange rates were also measured using NMR spectroscopy by following the intensities of the NMR peaks during deuterium exchange under similar conditions. Assignments can be made when the pK_a's or the exchange rates are sufficiently distinct. Data from NMR measurements and Ohe's 2D gel analysis technique provides a general method for the assignment of histidine NMR peaks in proteins.

Early NMR studies of staphylococcal nuclease resulted in tentative assignments of the histidine resonances by indirect techniques (15). Line H2 was assigned to His-46, the histidine known to be nearest the active site in the crystal structure, based on the effects of ligand binding on the NMR spectrum (16). Histidine 124 was unambiguously assigned by a comparative study of a naturally occurring nuclease variant, from *Staphylococcus aureus* strain V8, which differs only by a substitution of His-124 with a leucine residue (17).

The ease and certainty of the assignment of His-124 by comparison of the variant with wild type make it the preferred technique, but natural variants are rare and the only available point mutants at the other histidines disrupted enzyme activity or thermodynamic stability (18). We have engineered histidine substitution mutants in which His-46 or His-121 is replaced with asparagine. The asparagine residue was selected because it is roughly similar to histidine in gross shape and spatial distribution of polar atoms, and thus we hoped would not perturb the protein structure or stability. We present spectra and pH titration curves of the wild type nuclease; of the variant V8 which has His-124 ---> Leu (H124L); and of our engineered mutants His-121 ---> Asn (H121N) and His-46 ---> Asn (H46N)). These assignments are in agreement with those by Alexandrescu et al.(19) based on an analysis of an independent set of mutants.

METHODS

Expression of Nuclease in E. coli. A 525 bp Sau 3AI fragment encoding the nuclease A protein was isolated from a plasmid library of Staphylococcus aureus Foggi strain by antibody screening (20). The Sau3AI fragment was cloned into the BamHI site of the expression vector (21) pAS1. The resulting clone (pBF2) encoded a protein containing the sequence f-Met-Asp-Pro-Thr-Val-Tyr-Ser preceding the +1 codon of the mature nuclease A protein (22). The protein produced by this construct is designated nuclease R to indicate it is the result of the recombinant construct. The pBF2 construct produced about 70 mg/liter of nuclease.

The expression construct pBF2 was modified to remove the partial leader sequence by plasmid based oligonucleotide directed mutagenesis (23). DNA sequencing by the Sanger method confirmed that in the resulting expression construct pSNS1 the +1 codon for the nuclease A protein was fused to the initiating methionine codon of the pAS1 vector. The remainder of the sequence was identical to the wild type gene sequence. Protein sequencing has shown that the initiating methionine is cleaved from protein purified from this *E. coli* expression system.

Construction of histidine substitution mutants. Point mutants were generated by oligonucleotide directed mutagenesis. A 10-fold molar excess of mutagenic oligo was annealed to a single-stranded M13 DNA template with the nuclease gene. This was extended using Klenow fragment and transformed into JM101. Phage plaques were transferred to nitrocellulose filters to be probed by hybridization to the mutagenic oligo labelled with ^{32}P. Positive plaques were replated, screened by sequencing, and the modified nuclease gene subcloned into the pAS1 expression vector. H46N and H121N were generated by this method.

An expression construct for the mutant H124L was obtained by subcloning a DNA fragment of the H124L mutant gene (kindly provided by D. Shortle) from an internal HindIII site near codon 101 to the Sal I site 3' of the coding sequence into the similar sites of pBF2. Thus all of the substitution mutants in this paper are in the nuclease R background.

Growth of Bacterial Cultures, Induction, and Purification of Protein. The nuclease protein was prepared by growth of *E. coli* strain AR120 containing the pSNS1 plasmid or its mutant derivative in 12 l of 2xYT media (24) at 37°C with aeration and stirring in a fermentor (New Brunswick). The nuclease gene was induced by addition of 480 mg of nalidixic acid when the culture reached a density of OD_{600}=0.8. The culture was maintained for an additional 4 hours. Cells were harvested in a continuous flow rotor (Sorvall) and frozen at -20°C. The cell pellet was thawed in 10 mM ammonium acetate, pH 6.0, and passed twice through a French press. Cell debris was removed by centrifugation at 20,000xG and the supernatant, diluted with an equal volume of 0.6 M ammonium acetate, pH 6.0, was applied to a phosphocellulose column (Cellex-P, Bio Rad). The column was eluted with a 1:1 gradient of 0.3 M ammonium acetate, pH 6.0 and 1.0 M ammonium acetate, pH 8.0. Column fractions containing nuclease were pooled, diluted to be no more than 0.6 M salt, and passed over a DEAE column to remove nucleic acids which co-purified with nuclease. The effluent of this column was concentrated under N_2 at 60 psi in an Amicon concentrator, dialyzed twice against 4 l of 200 mM NaCl to remove the ammonium acetate buffer, and twice against distilled water. Several minor contaminating proteins precipitated during dialysis and were removed by centrifugation. The protein was lyophilized and stored at -20°C.

NMR Techniques. Lyophilized nuclease (100 mg) was suspended in 1 ml of D_2O (99.96%, Wilmad), adjusted to pH 5.5 with dilute NaOD and DCl in D_2O, then heated to its T_m for five minutes to facilitate the exchange of all amide protons. pH measurements were made in 1.5 ml microfuge tubes. Values of pH cited are uncorrected meter readings (25). Samples were lyophilized again and resuspended in 1 ml of 200 mM NaCl in D_2O, 20 µl of 30 mM sodium 3-trimethylsilylpropionate-2,2,3,3,-d_4 (TSP) was added, adjusted again to pH 5.5 and centrifugated for 15' in a microfuge before transferring to 5mm NMR tubes.

NMR spectra were recorded using the Pulse Fourier Transform method on a General Electric GN500 500 MHz spectrometer at Stanford University. At least 200 FID's of 16K points were accumulated allowing 6 sec (longest T1 = 3 sec) between 8 msec (60°) pulses, presaturating the residual water (HOD) peak for three seconds before acquisition. FID's were apodized using an exponential function equivalent to 2 Hz of line broadening before Fast Fourier Transformation. All chemical shifts cited are as ppm downfield from TSP.

All of a pH titration was done in a consecutive series, with the exception that additional low pH H121N data was gathered in a subsequent session. Samples were pH adjusted as above and the NMR tube was rinsed with adjusted sample before final measurement. Samples were allowed to equilibrate to temperature in the spectrometer for five minutes before shimming.

RESULTS AND DISCUSSION

NMR spectra for the nuclease R mutants, H46N, H121N, and H124L, collected at pH 5.5 in 200 mM acetate buffered D_2O, are shown in Figure 1. Each spectrum clearly shows the absence of one of the histidine C2 proton lines,

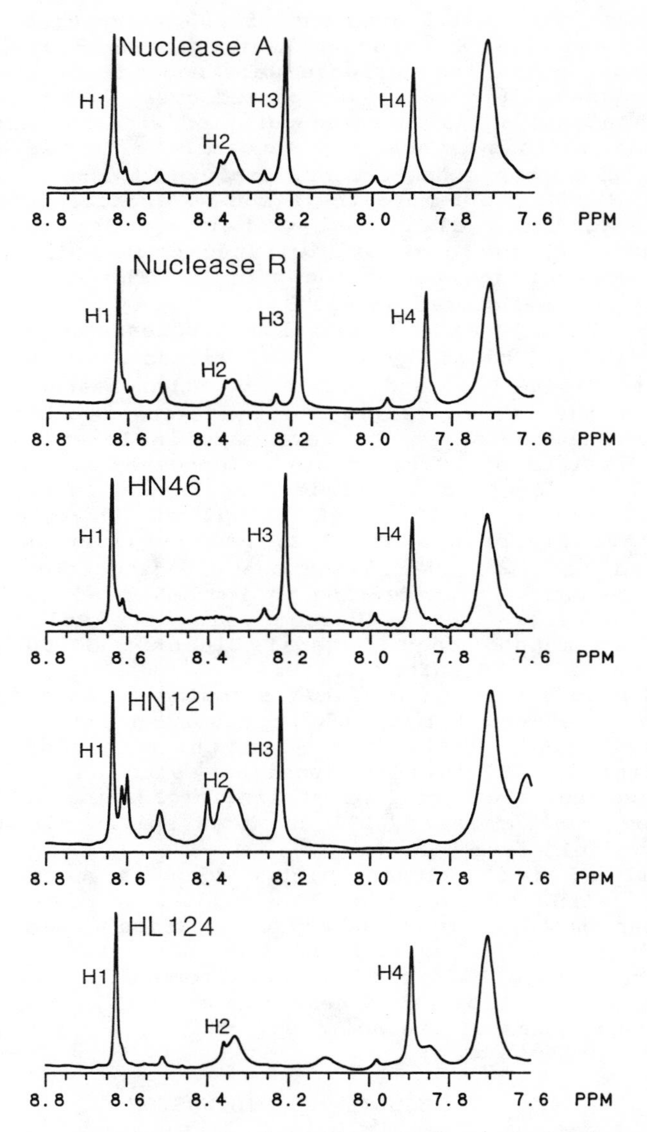

Figure 1: Spectra of histidine region of nuclease R and three histidine substitution mutants: H46N (His 46 to Asn), H121N (His 121 to Asn), and H124L (His 124 to Leu). Missing peaks show resonance assignments. Minor peaks at 7.97, 8.25, and 8.6 ppm are due to a stable structural isomer of nuclease.

leading to the assignments: H2=His-46, H3=His-124, H4=His-121, and by inference H1=His-8. The slight differences in chemical shift of the histidine C2 proton resonances between the wild type and the mutant proteins in these spectra may be due to effects of the substitutions, but are less than differences in chemical shifts observed for wild type protein in different salt conditions (data not shown). These chemical shift differences are not significant relative to the chemical shift separation of the four histidine C2 proton resonances in each spectrum.

The H1 resonance assigned to His-8 frequently displays two minor resonances. These are apparent in the nuclease A and nuclease R spectra in Figure 1, upfield from the major H1 peak. The more upfield minor peak is clearly resolved while the other is not fully resolved from the major resonance. The structural basis for one or both minor resonances may be distinct from the *cis:trans* isomerism which explains the minor resonances at H3 and H4 (6,8). The H121N mutant is of interest as the two minor resonances for H1 are much larger and better resolved than those seen in the wild type protein or in the other two mutants displayed in Figure 1. The sharp resonance near H2 at 8.40 ppm is a new minor histidine C2 proton resonance. At higher temperatures a minor resonance appears downfield from H3 at a chemical shift similar to that found in the wild type protein spectrum (data not shown).

The minor resonances at 7.95, 8.24, and 8.60 ppm are in chemical exchange with the major resonances (6). The observed chemical difference is due to the *cis/trans* heterogeneity of the lys116-pro117 peptide bond in the folded protein (8). In each of our substitution mutants one major and one minor resonance are absent, confirming the relationship indicated by two-dimensional magnetization transfer experiments (7). The H1 resonance is the only remaining major line in this region of the spectrum, and two-dimensional chemical exchange experiments of partially unfolded protein indicate that H1 is in chemical exchange with a resonance near the random coil chemical shift for the histidine C2 proton. The manner in which amino acid substitutions modulate this structural isomerization is the subject of ongoing studies (26). A small resonance appears in some samples at 8.47 ppm and is thought to arise from contaminating formate. The intensity of this resonance remains constant when the protein is thermally unfolded and does not titrate between pH 4.8 and 6.5, where histidine C2 protons display large chemical shift changes.

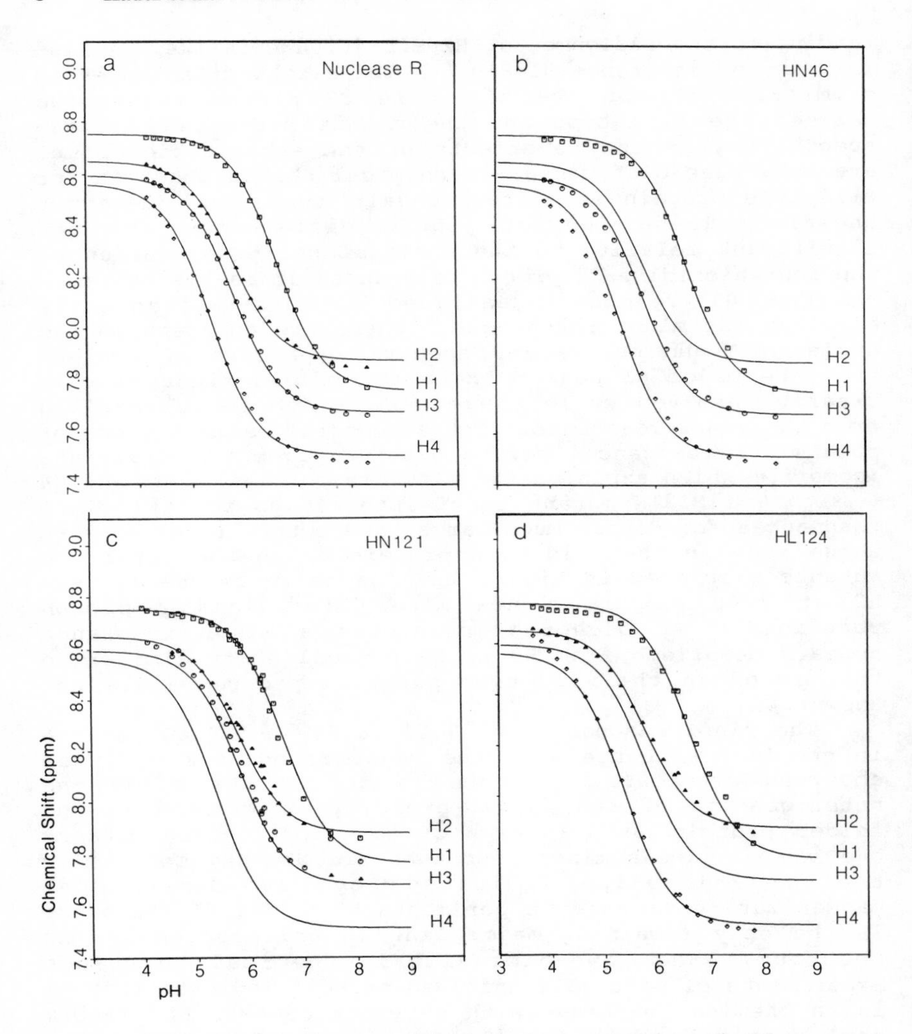

Figure 2: Histidine chemical shifts over a range of pH observed in a) nuclease R; b) H46N (His 46 to Asn); c) H121N (His 121 to Asn); d) H124L (His 124 to Leu). Samples are in 200 mM NaCl at 30°C, except H121N which was studied at 25°C because of its reduced thermal stability. Solid lines are a theoretical curve (text) fit by least-squares analysis to nuclease R chemical shift data.

A potential problem in making NMR resonance assignments by amino acid substitution is that the overall structure of the protein can be perturbed by the substitution, leading to chemical shift changes and thus ambiguities in the assignment (27). The chemical shift of the histidine C2 proton varies considerably with the titration of the imidizolium group. This behavior allows the assignments to be carried out at a variety of solution conditions, increasing the confidence in the assignments and also allowing a comparison of pK_a values in addition to chemical shifts. Although none of the histidine C2 proton resonances in Figure 1 shifted significantly, we have compared the spectra over a wide range of pH.

The chemical shift data from a pH titration of wild type nuclease R are shown in Figure 2a. The solid lines are least squares fits of a theoretical curve: The chemical shift of a histidine is a weighted average of its shift when it is completely protonated (its acidic asymptote) and when it is completely deprotonated (its basic asymptote). The weighting is given by the pH equilibrium. From the Henderson-Hasselbalch equation it can be derived:

$$\text{Chemical Shift} = \frac{\delta_a}{1+10^{(pH-pK)}} + \frac{\delta_b}{1+10^{(pK-pH)}}$$

where: δ_a = acid chemical shift, δ_b = basic chemical shift, pK = pK_a of the histidine. All three of δ_a, δ_b, and pK_a were varied to minimize the least squares fit. Panels b through d of Figure 2 show data from histidine substitution mutants H46N, H121N, and H124L. The solid lines are the curves fit to wild-type nuclease R superimposed on the mutant data for comparison. These data support the assignments derived from the spectra in Figure 1.

The theoretical curves fit the data reasonably well but there are two main discrepancies. The first is that at low pH the data points of H3 and H4 diverge downfield rather than approach the asymptote of the fitted curves. This divergence accompanies the appearance of the unfolded protein resonances at pH 4.2, indicating that the protein is entering the denaturational transition and so the histidines are no longer sampling only the folded protein environment. Secondly, the data points do not slope as steeply as the fitted curves near the midpoint of the titration. This could result from error in determining the low pH asymptote due to denaturation. Alternatively,

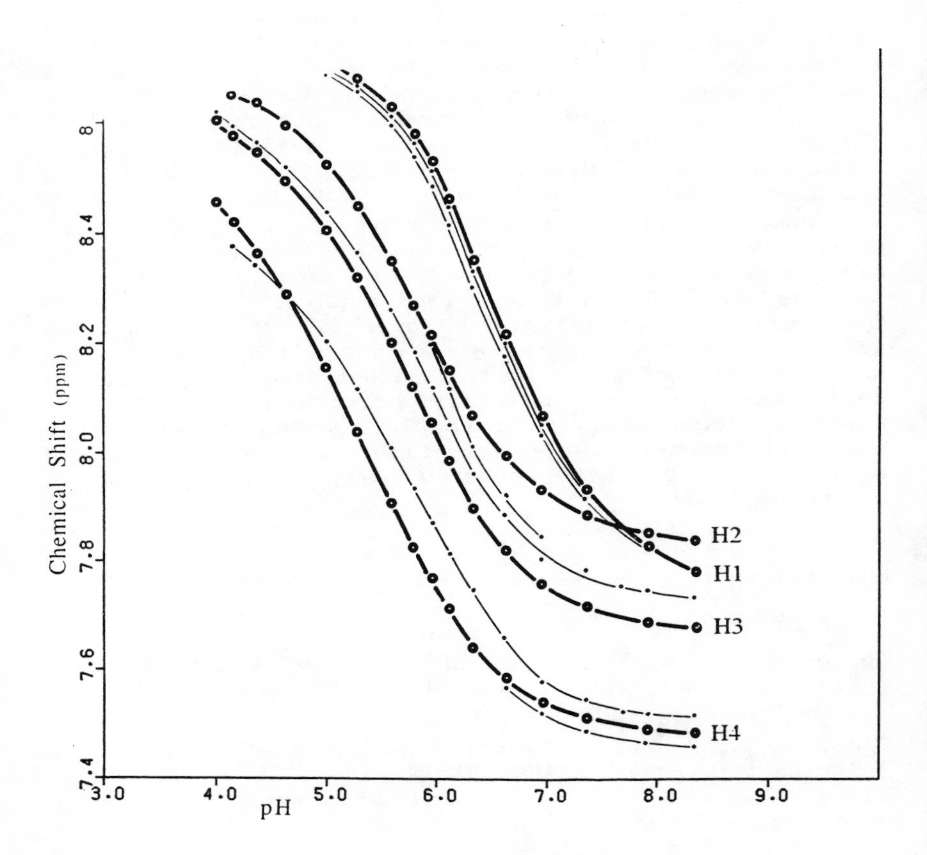

Figure 3: pH titration of nuclease A at 45° C. Thick lines connect data points of major resonances, thin lines show the appearance and position of minor resonances.

the histidine pK_a's may be pH dependent if they are influenced by charged groups which are also titrating in this pH range.

The titration curve of wild-type nuclease A is shown in more detail in Figure 3, including the appearance and chemical shifts of the minor peaks. The minor peaks follow the major peaks closely through the titration, strongly supporting the hypothesis that the minor peaks are histidine resonances. The minor peaks show the greatest separation from their associated major peaks between pH 5.3 and pH 5.5. Peak H4* appears to cross its major peak H4 at pH 4.3. The minor peaks H4*, H3*, and H1* have been studied extensively in association with isomerization of Pro-117. The other minor peaks have not been studied yet, and may be due to other isomerization events within nuclease. In particular, His-46, peak H2, is adjacent to Pro-47 which could be independently undergoing *cis/trans* isomerization.

The experiments presented here were performed on mutants in the nuclease R form, (which retains six of the leader peptide residues at its N terminus), because the histidine substitution mutants were analyzed before the nuclease A expression vector was constructed. The 5 N-terminal residues of nuclease A show disorder in the crystal structure (1,4), suggesting that these residues do not participate in the protein fold. Figure 1 shows the histidine C2 region of spectra of nuclease R and nuclease A. There are no significant differences in the chemical shift of any of the histidine resonances between the nuclease R and nuclease A spectra. Further, the melting temperatures of nuclease A and nuclease R are the same within the 0.5°C error of the determination. The nuclease R spectrum contains a number of sharp resonances, outside the histidine region, not found in the nuclease A spectrum. The sharpness of these resonances suggests the additional N-terminal residues are segmentally disordered.

Since the original work was done, mutants H124L and H121N have been produced in the A form. H124L(A) showed no differences from H124L(R) in T_m or in the histidine region of the NMR spectrum. H121N(A), however, has a T_m of 44°C, seven degrees higher than H121N(R). Also, the resonances of His-8 of H121N(A) are identical to those of wild type, while H121N(R) shows an additional minor peak H1**. The chemical shifts of all other histidine resonances agree closely.

TABLE 1.

pK_a's and T_m's of Point Mutants[a]

	NucR	H46N	H121N	H124L
H1 pK_a	6.6	6.6	6.6	6.6
H2 pK_a	5.8		5.9	5.8
H3 pK_a	5.8	5.7	5.8	
H4 pK_a	5.4	5.3		5.3
T_m (°C)	48.2	54.6	37.5	56.7

[a]T_m's and apparent pK_a's of wild-type and his substitution mutants. pK_a's were obtained by least squares fits of the data to a theoretical curve (text). T_m's were obtained from plots of ln(K) vs. 1/T (data not shown) of protein in 200 mM acetate, pH 5.5. Uncertainties of pK_a values are .03, except .05 for H121N; uncertainties of T_m's are 0.5°C.

Assigning histidine resonances in staphylococcal nuclease by comparing pH titrations of point mutants is unambiguous and relatively straightforward. The spectra of our substitution mutants were found to be virtually identical to wild type over the entire pH range except for the one histidine resonance absent from each substitution mutant. The pH titration curves show slight deviations from the theoretical curve of a histidine in isolation and may contain information concerning the interaction of the histidines with the electrostatic environment of the protein. The assignments made here will make it possible to interpret ongoing studies of the multiple conformations of nuclease in terms of the available high-resolution crystal structures. Nuclease is a particularly promising model system for studying the effects of specific site-directed mutations on the thermodynamic stability of proteins because it permits a quantitative comparison of two folded conformations in equilibrium, in addition to the equilibrium between the folded and unfolded states.

ACKNOWLEDGMENTS

We wish to thank Chris Dobson and Phil Evans for fruitful discussions and Thomas Hynes for providing information from the crystal structure. The 500 MHz NMR spectrometer at Stanford was obtained through NIH and NSF shared equipment grants. Roger Kautz is on leave from the Stanford Program in Biophysics.

REFERENCES

1. Cotton FA, Hazen EE.(1971). Staphylococcal Nuclease, X-ray Structure. In Boyer PD (ed): "The Enzymes, Vol IV. 3rd. edition". New York: Academic Press, p. 153.
2. Arnone A, Bier CJ, Cotton FA, Day VA, Hazen EE, Richardson DC, Richardson JS, Yonath AA (1971). High Resolution Structure of an Inhibitor Complex of the Extracellular Nuclease of Staphylococcus aureus I: Experimental Procedures and Chain Tracing. J. Biol. Chem, 246:2302-2316.
3. Cotton FA, Bier CJ, Day VW, Hazen EE, Larsen S (1972). Some Aspects of the Structure of Staphyloccal Nuclease. In "Cold Spring Harbor Symposia on Quantitative Biology, Volume XXXVI Structure and Function of Proteins at the Three Dimensional Level". New York: Cold Spring Harbor Laboratory, pp. 243-255.
4. Hynes TR et. al., in prep
5. Loll PJ, Lattman EE (1989). The Crystal Structure of the Ternary Complex of Staphylococcal Nuclease, CA^{2+}, and the Inhibitor pdTp, Refined at 1.65 A (submitted).
6. Fox RO, Evans PA, Dobson CM (1986). Multiple Conformations of a Protein Demonstrated by Magnetization Transfer NMR Spectroscopy. Nature 320:192-194.
7. Evans PA, Kautz RA, Fox RO, Dobson CM (1988). A Magnetisation Transfer NMR Study of the Folding of Staphylococcal Nuclease. Biochem (in press).
8. Evans PA, Dobson CM, Kautz RA, Hatfull G, Fox RO (1987). Proline Isomerization in Staphylococcal Nuclease Characterized by NMR and Site-Directed Mutagenesis. Nature 329:266-268.
9. Dalvit C, Wright P (1987). Assignment of Resonances in the ^{1}H Nuclear Magnetic Resonance Spectrum of the Carbon Monoxide Complex of Sperm Whale Myoglobin by Phase-sensitive Two-dimensional Techniques. J. Mol. Biol. 194:313-327.
10. Kline AD, Wuthrich K (1986). Complete Sequence-Specific ^{1}H Nuclear Magnetic Resonance Assignments For The a-Amylase Polypeptide Inhibitor Tendamistat

From Streptomyces tendae. J. Mol. Biol. 192:869-890.

11. Lemaster DM, Richards FM (1988). NMR Sequential Assignment of *Escherichia coli* Thioredoxin Utilizing Random Fractional Deuteration. Biochem 27:142-150.
12. Redfield C, Dobson CM (1988). Sequential ^{1}H NMR Assignments and Secondary Structure of Hen Egg White Lysozyme in Solution. Biochem 27:122-136
13. Meadows DH, Jardetzky O, Epand RM, Ruterjans HH, Scheraga,HA (1968). Assignment of the Histidine Peaks in the Nuclear Magnetic Resonance Spectrum of Ribonuclease. PNAS 60:766-722.
14. Ohe M, Matsuo,H, Sakiyama F, Narita K (1974). Determination of pKa's of Individual Histidine Residues in Pancreatic Ribonuclease by Hydrogen-Tritium Exchange. J. Biochem (Tokyo) 74:1197-1200.
15. Jardetzky O. Theilmann H, Arata Y, Markley JL, Williams MN (1972). Tentative Sequential Model for the Unfolding and Refolding of Staphylococcal Nuclease at High pH. In "Cold Spring Harbor Symposia on Quantitative Biology, Volume XXXVI Structure and Function of Proteins at the Three Dimensional Level". New York: Cold Spring Harbor Laboratory, pp. 257-261.
16. Epstein HF, Schechter AN, Cohen JS (1971). Folding of Staphylococcal Nuclease: Magnetic Resonance and Fluorescence Studies of Individual Residues. PNAS 68:2042-2046.
17. Markley JL (1969) PhD Thesis, Harvard University, Cambridge Mass. (reprinted in Jardetzky O, Roberts GCK (1981) "NMR in Molecular Biology". Academic Press, Inc.)
18. Shortle D, Lin B (1985). Genetic Analysis of Staphylococcal Nuclease: Identification of Three Intragenic "Global" Suppressors of Nuclease-Minus Mutations. Genetics 110:539-555.
19. Alexandrescu AT, Mills DA, Ulrich EL, Chinami M, Markley JL (1988). NMR Assignments of the Four Histidines of Staphlococcal Nuclease in Native and Denatured States. Biochem 27:2158-2165.
20. Fox RO, unpublished results.
21. Rosenberg M, Ho YS, Shatzman A (1983). The Use of pKC30 and Its Derivatives for Controlled Expression of Genes. In "Methods in Enzymology, Vol 101". Academic press, Inc.
22. Davis A, Moore IB, Parker DS, Taniuchi H (1977). Nuclease B: A Possible Precursor of Nuclease A, An Extracellular Nuclease of Staphylococcus Aureus. J Biol Chem 252:6544-6553.

23. Inouye S, Inouye M (1986). Oligonucleotide-directed Site-specific Mutagenesis Using Double-stranded Plasmid DNA in DNA and RNA synthesis (ed. S. Narang) Academic Press
24. Maniatis T, Fritsch EF, Sambrook J (1982). "Molecular Cloning: A Laboratory Manual" New York: Cold Spring Harbor Laboratory.
25. Bundi A. Wuthrich K (1979). ^{1}H-NMR Parameters of the Common Amino Acid Residues Measured in Aqueous Solutions of the Linear Tetrapeptides H-Gly-Gly-X-Ala-OH. Biopolymers 18:285-297.
26. Kautz RA, Fox RO, in prep.
27. Hibler D, Stolowich N, Reynolds H, Gerlt J, Wilde J, Bolton P (1987). Site-Directed Mutants of Staphylococcal Nuclease: Detection and Localization by ^{1}H NMR Spectroscopy of Conformational Changes Accompanying Substitutions for Glutamic Acid-43. Biochem 26:6278-6286.

Protein and Pharmaceutical Engineering, pages 17–33

DETECTION OF CONFORMATIONAL CHANGES IN ACTIVE SITE MUTANTS OF STAPHYLOCOCCAL NUCLEASE[1]

David W. Hilber, Tayebeh Pourmotabbed,
Mark Dell'Acqua and John A. Gerlt

Department of Chemistry and Biochemistry
University of Maryland, College Park, MD 20742

Susan M. Stanczyk and Philip H. Bolton

Department of Chemistry, Wesleyan University
Middletown, CT 06457

Patrick Loll and Eaton Lattman

Department of Biophysics, Johns Hopkings University
School of Medicine, Baltimore, MD 21205

ABSTRACT We have generated site-directed mutations of Glu 43, Arg 35, and Arg 87 in the active site of Staphylococcal nuclease (SNase). On the basis of the newly refined 1.65 Å x-ray structure, Glu 43 is proposed to act as a general base catalyst and both Arg 35 and Arg 87 are proposed to act as electrophilic catalysts in the attack of water on the phosphodiester bond of the substrate. All of the substitutions made for these residues significantly decrease the catalytic efficiency but do not totally inactive the enzyme. High resolution ^{1}H NMR studies suggest that all of the substitutions we have made produce conformational changes within the active site that are propagated into the hydrophobic core and to the environments of histidine residues remote from the active site. Finally, the structure of the E43D mutant has been solved to 1.8 Å resolution, and this

---

[1]This was was supported by NIH GM-36358 (to E.L.) and NIH GM-34573 (to J.A.G. and P.H.B.)

reveals the presence of conformational changes both within and remote from the active site.

INTRODUCTION

Staphylococcal nuclease (SNase) is a monomeric protein composed of 149 amino acids in a known sequence determined both by chemical degradation (1) and DNA sequencing techniques (2). The enzyme is synthesized by _Staphylococcus aureus_ as a preenzyme which is secreted across the cell membrane into the extraceullar medium concomitant with cleavage of a leader peptide. SNase catalyzes the hydrolysis of single stranded DNA and RNA only in the presence of Ca^{2+} to yield 3'-mononucleotides as products. The rate of hydrolysis of DNA catalyzed by SNase exceeds the uncatalyzed rate by a a factor of approximately 10^{15}. A highly refined 1.65 Å x-ray structure determined in the presence of Ca^{+2} and the competitive inhibitor thymidine 3',5'-bisphosphate (pdTp) was recently completed (3); this structure differs only slightly from a less refined 1.5 Å structure that was reported in 1979 (4). The x-ray structure shows the positions of amino acids in the active site (Figure 1) that can be used to formulate a mechanism. The Ca^{2+} ion, held in place by coordination to the β-carboxylates of Asp 21 and Asp 40 and to the carbonyl oxygen of Thr 41, is coordinated to the 5'-phosphate group of the competitive inhibitor. This phosphate group is also coordinated to the guanidinium functional groups of Arg 35 and Arg 87. If the assumption is made that the position of the bound inhibitor mimics the position of bound substrate and since SNase produces 3'-mononucleotides, the position of the scissile P-O bond is between the 5'-phosphate group and its 5'-ester oxygen. Thus, the Ca^{2+} ion and the guanidinium groups of both arginines can be proposed to be electrophilic catalysts that facilitate the direct attack of water on the phosphorus to effect P-O bond cleavage. The γ-carboxylate group of Glu 43 is also present in the vicinity of the 5'-phosphate, and in the x-ray structure it is hydrogen bonded to two water molecules, one which is in the inner coordination sphere of the Ca^{2+} and a second which serves as a bridge between the carboxylate group and the 5'-phosphate group of the bound inhibitor. Thus, the γ-carboxylate group of Glu 43 can be proposed to be a general basic catalyst that facilitates the direct attack of water on the phosphorus. The stereochemical course of

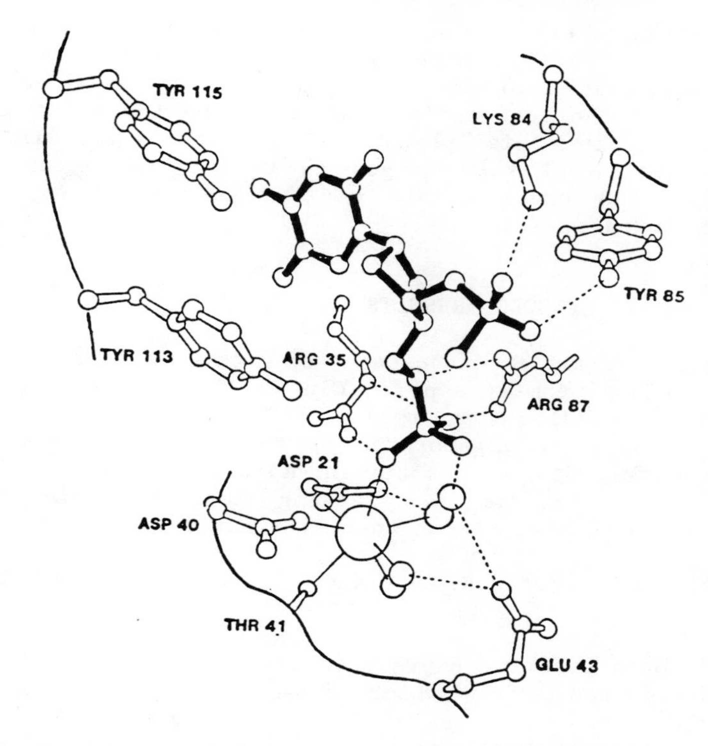

FIGURE 1. Active site of Staphylococcal nuclease.

the reaction, inversion, is in accord with this proposal (5).

[We note that the x-ray structure also reveals an intermolecular contact between two SNase molecules in the crystal which, unfortunately, directly involves the γ-carboxylate group of Glu 43. The ϵ-ammonium group of Lys 70 in the adjacent SNase molecule forms a salt bridge with the carboxylate group of Glu 43. The effect of this interaction on the hypothesis concerning mechanism is uncertain but is being tested by the construction of site-directed mutants which will eliminate this salt bridge.]

We have sought to test these proposed roles for Glu 43, Arg 35, and Arg 87 by the construction of specific mutants by site-directed mutagenesis. In addition to measurement of the kinetic parameters for the mutant enzymes, we have focussed considerable attention on determining whether the conformations of the mutant proteins differ from that of wild type SNase. We have concluded that conformational changes both within and remote from the active site are caused by the amino acid substitutions. This phenomenon prevents quantitative interpretation of the changes in kinetic parameters in terms of the precise role of the mutated amino acid residue in catalysis.

METHODS, RESULTS, AND DISCUSSION

This article focusses on three conservative substitutions, namely Asp for Glu 43 (E43D), Lys for Arg 35 (R35K), and Lys for Arg 87 (R87K). While additional mutations have been generated at these sites, e.g., Gln and Asn for Glu 43, most enzymologists would review retention of charge as creating less potential for conformational alterations than elimination of charge.

Site-Directed Mutagenesis and Expression of Mutant Proteins

We have used two expression plasmids in our studies, one designated pONF1 which directs the secretion of properly processed SNase into the periplasmic space of the *Escherichia coli* host and (6) the second designated pNJS which leads to accumulation of large amounts of intracellular SNase (7). The first plasmid was constructed by fusion of the coding sequence for the 149 amino acid protein secreted by *S. aureus* to the 3'-end of the coding sequence for the leader peptide associated with the *omp*A protein in *E. coli*. This gene is downstream of a *lac* promoter so that induction can be accomplished by the addition of lactose or IPTG. With this system approximately 10% of the cell protein accumulates as the processed SNase. The second plasmid was constructed by ligation of a *Sau* 3A restriction fragment containing the gene into the *Bam* HI site of the expression vector pCQV2. In this plasmid, the gene for the nuclease is extended by seven codons which are present at the 5'-flanking end of the structural gene for the processed protein in the *Sau* 3A fragment. This gene is downstream of the bacteriophage λ P_R promoter so its regulation can be controlled in a cell in which the cI857 temperature sensitive λ repressor is produced. With this system approximately 50% of the cell protein accumulates as the N-terminal extended SNase. Plasmid pONF1 was used for the construction, enzymological characterization, and crystallization of E43D, and pNJS was used for all of our studies of R35K and R87K.

The proteins produced by both expression plasmids can be readily purified by chromatography on BioRex-70 cation exchange columns followed by affinity chromatography on pdTp derivatized Sepharose.

The procedures we have used for mutageneses have evolved with the available M13 based methods: E43D was constructed according to a double primer method (8), and both R35K and R87K were constructed according to the method employing a deoxyuridine containing template (9).

Kinetic Characterization of Mutants

The purified mutant proteins have been subjected to kinetic characterization at pH 9.5, the pH optimum of wild type SNase, and in the presence of 10 mM Ca^{2+}, saturating for the mutants as well as wild type SNase, so that the effect of the amino acid substitutions on catalytic efficiency could be assessed. The kinetic parameters that were obtained are summarized in Table I.

TABLE I.
KINETIC CHARACTERIZATION OF WILD TYPE AND MUTANT SNASES.

Protein	relative V_{max}	K_m [a]	relative V_{max}/K_m
Wild type	1000.0	20	1000.0
E43D	5.2	115	0.90
R35K	0.48	73	0.14
R87K	ND[b]	ND[b]	0.08[b]

[a] μg/mL of DNA
The K_m was too high to measure so neither K_m nor V_{max} could be determined accurately; the relative V_{max}/K_m was determined by comparing velocities at low substrate concentrations.

Clearly, each conservative substitution causes a significant decrease in catalytic efficiency, as assessed by the ratio V_{max}/K_m. The fact that the K_m values for the mutant enzymes differ from that of wild type SNase demonstrates that the decreased catalytic efficiencies can not simply be attributed to contamination by small amounts of wild type SNase. The decreases in catalytic activity that are produced by the substitutions may be viewed as large but they are small relative to the 10^{15} rate acceleration that is characteristic of wild type SNase. If possible conformational changes are ignored, one interpretation of these data is that general base catalysis by Glu 43 is responsible for a factor of 10^3 and that electrophilic catalysis by Arg 35 and Arg 87 is each responsible for a factor of 10^4 in the total rate acceleration. Thus, it could appear that the majority of the rate acceleration is contributed by these three functional groups and that rather little is left for electrophilic catalysis by Ca^{2+} and for entropic effects.

Stability Characterization of Mutants

SNase has been demonstrated to be reversibly denatured by heat, and the thermodynamic parameters for thermal denaturation of wild type SNase have been reported (10). Although the melting temperature (T_m) of a protein cannot be used to rigorously compare the stabilities of proteins, the determination of the T_m does provide a convenient, although qualitative, way to assess whether amino acid substitutions in the active site are likely to alter the stability of the mutant SNase. The T_m values were determined for wild type and mutant SNases both in the presence and absence of active site ligands (Ca^{2+} and pdTp), and these are collected in Table 2.

The T_m of wild type SNase is increased by the presence of active site ligands, in accord with the calorimetric measurements made by Sturtevant (10). Note that in the absence of active site ligands E43D and R35K have T_ms higher than wild type; in the presence of ligands the situation is reversed. The same observation has been made for several other active site mutants. The melting behavior of R87K is anomalous since it is the only mutant we have studied which has a T_m lower than wild type in the absence of ligands. Since R87K does not have a high affinity for pdTp, the precise meaning of the T_m determined in the presence of ligands is questionable. In any event, the T_ms for the mutant SNases differ from that of wild type SNase, and these differences suggest that the

TABLE 2.
MELTING TEMPERATURES OF WILD TYPE AND MUTANT SNASES.

Protein	T_m - Ligands[a]	T_m + Ligands[b]
Wild Type	52.0 ^oC	62.4 ^oC
E43D	53.0	56.9
R35K	56.2	58.7
R87K	47.6	52.6

[a]0.1 M NaCl, 25 mM Borate, pH 7.8.
[b]Same as a) plus 10 mM $CaCl_2$ and 1×10^{-5} M pdTp.

stability of either the folded or the unfolded form of the enzyme (or both) is altered by the substitutions in the active site. Since stability differences are likely to arise from structural changes, these data may suggest that the substitutions alter the structure of the folded forms.

If the structures of the active sites of the mutant enzymes do differ from that of wild type, the simplest interpretation of the decreased catalytic efficiences made in the previous section are likely to be in error. Small changes in geometric disposition of the substrate and functional groups in the active site are likely to produce large changes in reaction velocity, especially given the fact that the SNase is such an efficient catalyst. In this situation, quantitative description of the role of a given active site functional group in catalysis would be impossible.

^{1}H NMR Characterization of Mutants

In view of the suggestion that the active site substitutions alter the structure of the folded forms of the mutant enzymes relative to that of wild type and the anticipation that such changes could prevent interpretation of the changes in kinetic parameters that have been measured, we have simultaneously pursued both ^{1}H NMR spectroscopy and x-ray crystallography to assess whether conformational changes are, in fact, present in the mutant enzymes. Although Dr. Dennis Torchia's group at NIH in Bethesda, Maryland recently has accomplished the total assignment of the two-dimensional ^{1}H chemical shift correlation NMR spectrum of wild type SNase, he has not yet used this information to completely describe the folded form of wild type SNase. Nor has he yet investigated any of the active site mutants described in this chapter. Thus, we have directed our attention to using one-dimensional and limited two-dimensional ^{1}H NMR spectroscopy as qualitative methods for comparing the structures of mutant and wild type SNases. Clearly, our use of ^{1}H NMR spectroscopy will provide only a limited amount of structural data in comparison to that which ultimately will be obtained ^{1}H NMR spectroscpoy or that which already has been obtained by x-ray crystallography but it does provide a rapid method to make qualitative comparisons.

Comparison of E43D with Wild Type We recently described the results of some of our ^{1}H NMR studies of E43D, and these will not be described in detail in this article (7,11). To summarize these results we note that changes in the one-dimensional ^{1}H NMR spectra were observed both in the reasonably well resolved aromatic and upfield shifted proton regions. By deuteriation of aromatic and aliphatic amino acids using our expression plasmid pNJS, we were able to assign the resonances experiencing chemical shift changes to Leu 25, Tyr 27, Phe 34, Val 74, Phe 76, Ile 139, and Trp 140. The first five residues are located in or near the hydrophobic core approximately 25 Å from residue 43; the last two residues are located on the surface approximately 12 Å from residue 43. Many other chemical shift changes are noted in the unresolved regions of the spectrum. At least some of these assigned resonances also experience changes in the intensities of interresidue nuclear Overhauser effects, so the changes in chemical shift are certainly caused by a changes in conformation. That the conformational changes are detected considerable distances from residue 43 suggests that the replacement of glutamate with aspartate causes a significant change in structure in the active site which is propagated through the structure of the protein, and such changes makes precise interpretation of the measured decrease in catalytic efficiency impossible.

Comparison of R35K and R87K with Wild Type We have recently compared the one-dimensional ^{1}H NMR spectra of R35K, R87K, and wild type SNase. Representative data which summarize our conclusions are shown in Figures 2 and 3. These figures compare the upfield shift proton regions

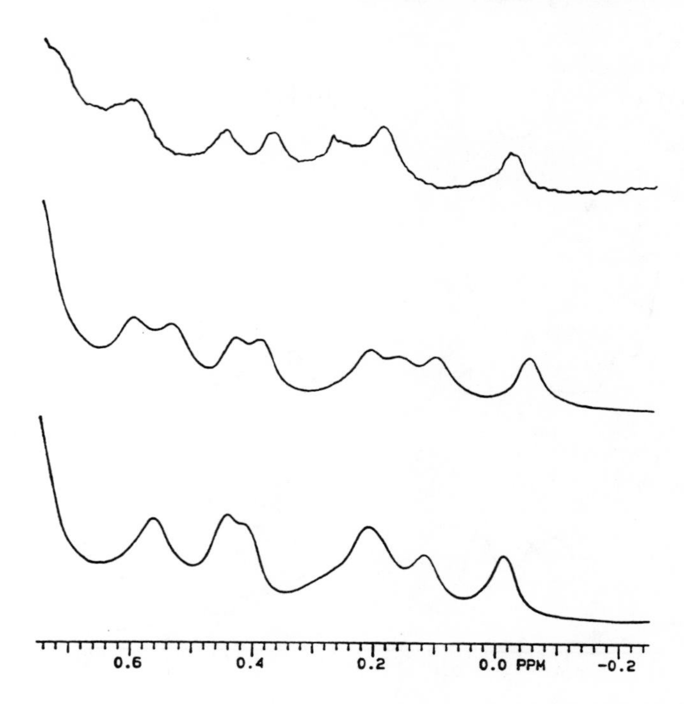

FIGURE 2. ^{1}H NMR spectra of R87K (top), wild type (middle), and R35K (bottom) in the absence of ligands.

of the spectra of these three proteins in the absence and presence of active site ligands, respectively. In both situations, the spectra of the proteins containing the homologous substitutions for the arginine residues differ from that of wild type, with the changes involving the same residues that were identified in our studies of E43D. In obtaining these spectra we first added Ca^{2+} to the unliganded samples, and in the case of wild type and R35K but not in the case of R87K significant changes in chemical shift were observed. When pdTp was then added to the samples, again only wild type and R35K experienced significant changes in chemical shift. Our conclusions based upon the data in the figures and the changes observed upon adding ligands is that both R35K and R87K differ in conformation from wild type, with R87K being the more severely "damaged" since it does not appear to bind the active site ligands. Thus, we anticipate that in both

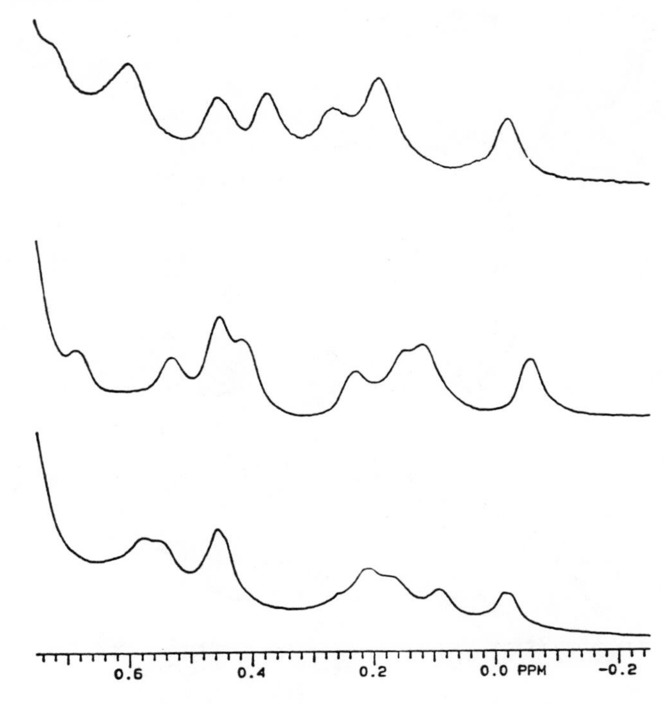

FIGURE 3. ^{1}H NMR spectra of R87K (top), wild type (middle), and R35K (bottom) in the presence of ligands.

mutants a significant change in structure occurs in the active site which is propagated through the structure of the protein. Again, such changes make precise interpretation of the measured decreases in catalytic efficiency impossible.

Effect of Active Site Mutations on Histidine Resonances We recently reported our use of deuteriation of the phenylalanine, tyrosine, and tryptophan residues in SNase to allow observation of the H_δ as well as H_ϵ resonances (12). In the ^{1}H NMR spectra of most proteins the H_δ resonances are obscured by the aromatic resonances, and, accordingly, these have been rarely studied. Much to our surprise, two resonances were found for the H_δ proton of His 121. A single resonance is observed for the H_ϵ proton of His 121, and two resonances are observed for each of the remaining three histidine resonances. We hypothesize that the unexpected behavior of the resonances for His 121 may be explained by the fact that the rotamers about the β,γ-bond of the side chain are chemically distinct because of the separate hydrogen bond donating and accepting properties of the nitrogens; upon rotation about this bond the H_δ proton experiences a large change in environment whereas the H_ϵ proton is minimally affected. We assume that two resonances are observed for the H_δ proton of His 121 because in one conformation it acts as a hydrogen bond donor to a neighboring side chain and in the second conformation it acts as a hydrogen bond acceptor to another residue.

Due to this unusual behavior of His 121, we thought that the effect of amino acid substitutions in the active site on the histidine resonances might be useful in obtaining additional evidence for global conformational changes. Samples of E43D, R35K, R87K, P117G, and wild type SNase were prepared in which the phenylalanine, tyrosine, and tryptophan residues were deuteriated. The P117G mutant SNase was obtained from Professor Robert O. Fox, Yale University. This mutation was constructed to test the hypothesis that *cis-trans* isomerism about the Lys 116-Pro 117 peptide bond might be the cause of multiple and minor resonances for the H_ϵ resonances of the histidine residues (13,14). According to the x-ray structure, this peptide bond is *cis*, and in solution Fox and Dobson have proposed that slow isomerism can occur such that approximately 10% of the molecules have the *trans* conformation. In accord with this hypothesis, the minor resonances are missing in the ^1H NMR spectrum of P117G.

The spectra we obtained are reproduced in Figures 4 and 5. Figure 4 shows the spectra of R87K, wild type, and R35K, from top to bottom; Figure 5 shows the spectra of E43D, wild type, and P117G, from top to bottom. The first feature to be noted is that in each spectrum multiple resonances for H_δ of His 121 occur at 6.9 ppm. The contaminating protiated amino acids prevents detection of minor changes in the chemical shifts or intensities of these resonances; however, it is clear that no major changes occur in the resonances associated with H_δ of His 121.

The second feature to be noted is that the intensities of the minor resonances for the *trans* isomer of the Lys 116-Pro 117 peptide bond which are clearly visible between the resonances for the H_ϵ protons of His

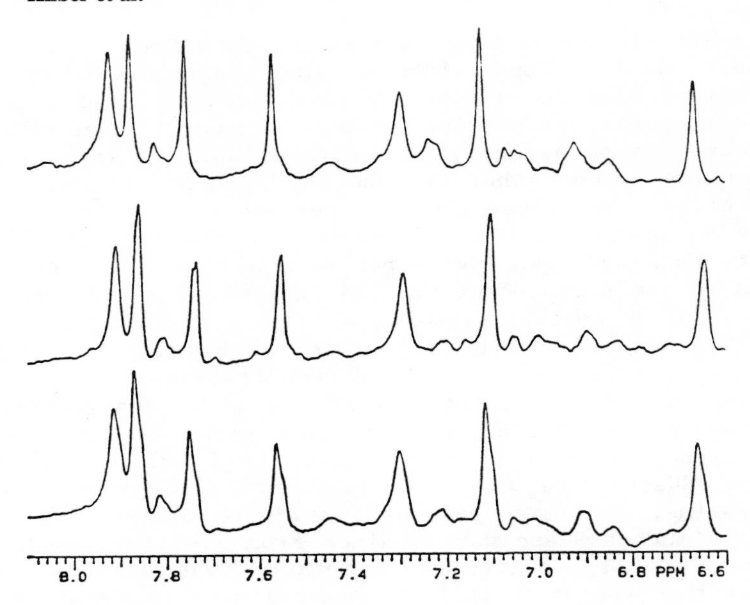

FIGURE 4. ^{1}H NMR spectra of R87K (top), wild type (middle), and R35K (bottom).

121 (7.56 ppm) and His 124 (7.76 ppm) are altered in intensity in the various mutants (the small resonance upfield of the resonance for H_ϵ of His 124 is the minor resonance for His 124, and the small resonance downfield of the resonance for H_ϵ of His 121 is the minor resonance for His 121). As expected, the resonances are missing from P117G, but they are also missing from the spectra of R35K and R87K. In the spectrum of E43D they are increased in intensity. Thus, substitutions within the active site appear to alter the equilibrium constant for the *cis* and *trans* isomers of the prolyl peptide bond. Since Pro 117 and His 121 are each approximately 25 Å removed from Glu 43, the alterations in the intensities of the minor resonances must be the result of the global conformational changes induced by the active sites substitutions that were previously detected by ^{1}H NMR in two other regions of the structure.

Thus, ^{1}H NMR spectroscopy aided by the ability to prepare samples of wild type and mutant SNases that are deuteriated in specific amino acid types has permitted the detection of conformational changes distant from the

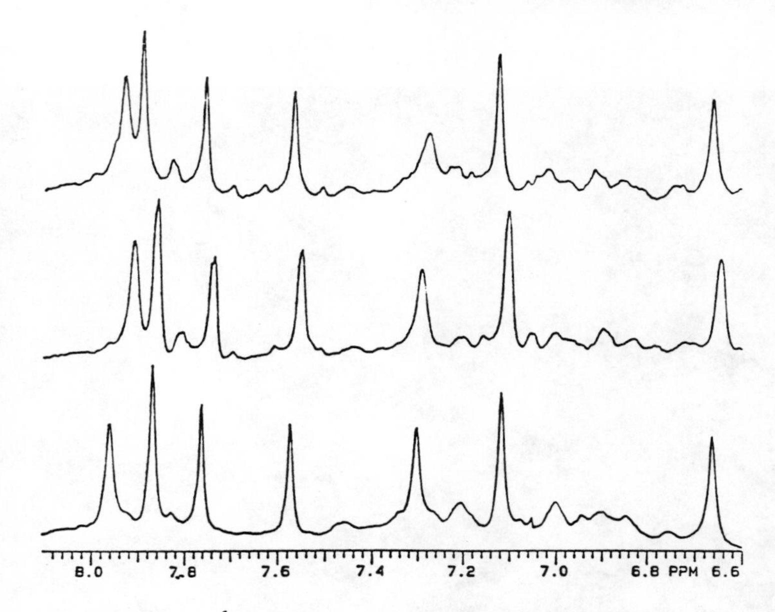

FIGURE 5. ^{1}H NMR spectra of E43D (top), wild type (middle), and P117G (bottom).

active site. Apparently, the substitutions we have made do alter the structure of the active site and this alteration is propagated throughout the molecule. Quantitative interpretation of the observed decreases in catalytic efficiency is not possible.

X-Ray Characterization of E43D

As previously noted, the structure of wild type SNase was recently refined to 1.65 Å resolution with a crystallographic R-factor of 0.161. The structure of E43D has been refined to 1.8 Å resolution with an R-factor of 0.186. The availability of both of these structures allows direct observation of the conformational changes produced by the substitution of aspartate for Glu 43.

The polypeptide backbones of the two proteins are nearly superimposable with the exception of a flexible loop near the active site (Figure 6). The electron density which defines this loop is ill-defined in both structures due to its flexibilty, and considerable uncertainty does exist in this region. However, in accord

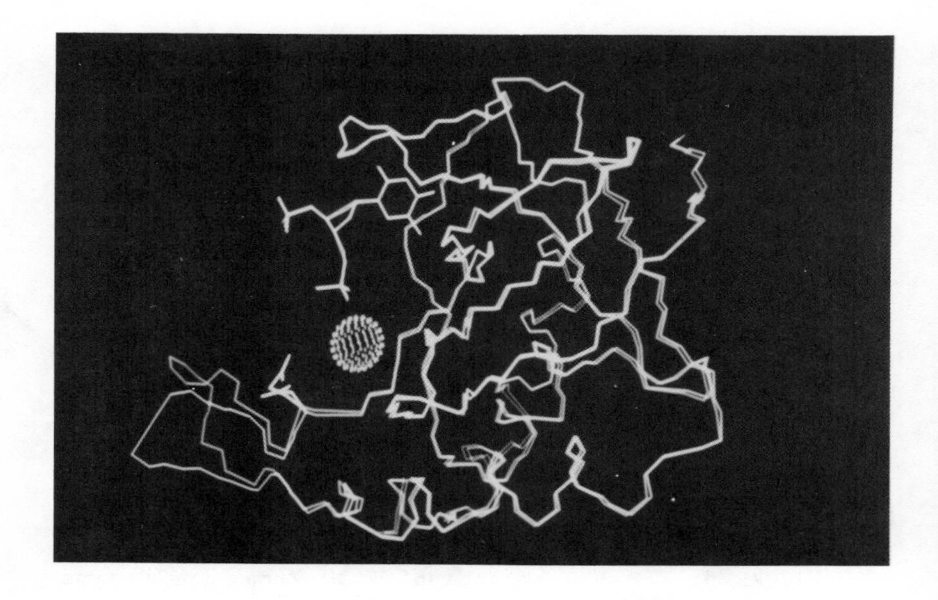

FIGURE 6. Backbones of wild type and E43D.

with the expectations based upon the thermal stability and ^{1}H NMR spectroscopy, superpositioning of the structures for wild type and E43D SNase reveals significant changes in the positions of side chains both within the active site and well removed from the active site.

The structures of the active sites are compared in Figure 7. In wild type SNase, Glu 43 is hydrogen bonded to a water coordinated to the Ca^{2+}. In E43D Asp 43 cannot reach this water molecule, and the side chain is rotated 180° around χ_1 such that the carboxylate group is hydrogen bonded to another water molecule that is not coordinated to the Ca^{2+}. Perhaps in response to this change in the position of a negative charge, Glu 52 moves toward the Ca^{2+}. In both proteins the Ca^{2+} is heptacoordinate in an octahedral geometry. However, in E43D the metal ion ligands have moved away from the Ca^{2+} such that its coordination sphere is "exploded." This movement apparently causes a change in the conformation of Ile 18 and Val 23 which are located on the β-turn which includes

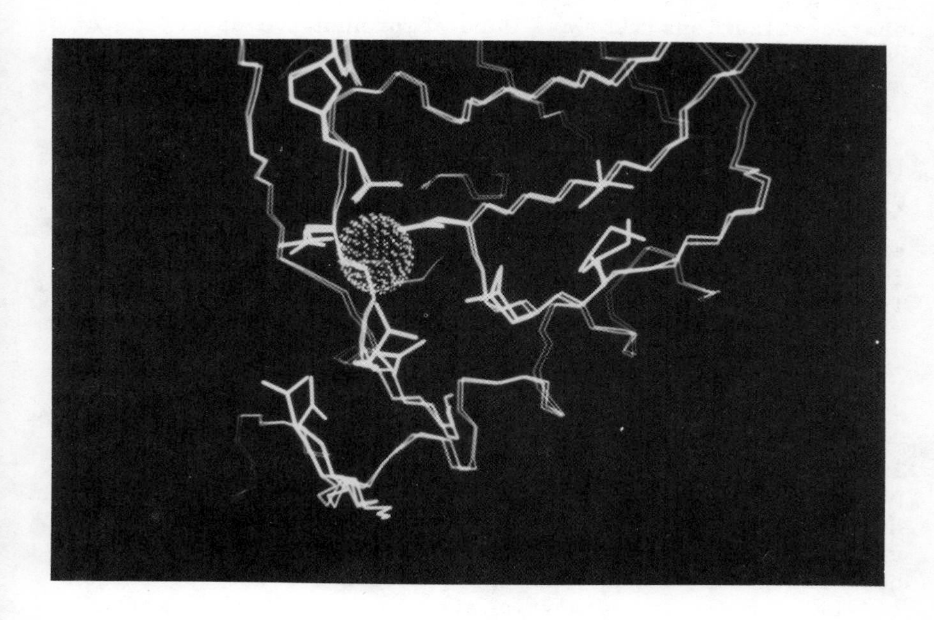

FIGURE 7. active sites of wild type and E43D.

one direct Ca^{2+} ligand (Asp 21) and one indirect Ca^{2+} ligand (Asp 19 is hydrogen bonded to a coordinated water). Several other changes some distance away from the active site are observed, including the positions of the side chains of Phe 61 and Thr 62.

No crystallographically detectable conformational changes are noted in the hydrophobic core and in the vicinity of Pro 117, His 121, Ile 139, and Trp 140, regions that ^1H NMR disclosed as experiencing conformational alterations. However, at least in the case of the hydrophobic core it is interesting to note that Val 23 which is rotated 120° about its χ_1 is located within the hydrophobic core and is spatially proximal to the residues implicated by ^1H NMR. Thus, a "pathway" for transmission of the conformational perturbation can be proposed. Whether the changes detected by ^1H NMR and the absence of changes in these regions as assessed by x-ray crystallography should be interpretted as structural differences between solution and crystal structures is uncertain. Certainly, the use of both types of structural characterizations allows a more comprehensive detection of conformational changes.

In any event, the expectation that mechanistically significant conformational changes would be found in E43D was confirmed by x-ray crystallography. Given the fact that changes were observed within the active site and, in particular, in the coordination sphere of the essential Ca^{2+}, the interpretation of decreases in catalytic efficiency as being quantitative measures of the importance of each mutated residue in catalysis is clearly incorrect.

SUMMARY

Conservative active site substitutions have been made in the active site of Staphylococcal nuclease. All of the mutant enzymes have decreased catalytic efficiency. However, by all criteria that have been applied conformational changes accompany the substitutions. These conformational changes prevent quantitative interpretation of the changes in catalytic efficiency. Whether methods can be devised to counter or prevent the conformational changes so that mechanistically useful information might eventually be obtained remains to be tested.

REFERENCES

1. Bohnert JL, Taniuchi H (1972). The examination of the presence of amide groups in glutamic and aspartic acid residues of Staphylococcal nuclease. J Biol Chem 247:4557.
2. Shortle D (1983). A genetic system for analysis of Staphylococcal nuclease. Gene 22:181.
3. Loll P, Lattman EE, (1989). The crystal structure of the ternary complex of Staphylococcal nuclease, Ca^{2+}, and the inhibitor pdTp refined at 1.65 Å. Proteins, in press.
4. Cotton FA, Hazen EE, Legg MJ (1979). Staphylococcal nuclease: proposed mechanism of action based on structure of enzyme-thymidine 3',5'-bisphosphate-calcium ion complex at 1.5 Å resolution. Proc Nat Acad Sci USA 76:2551.

5. Mehdi S, Gerlt JA (1982). Oxygen chiral phosphodiesters. 7. Stereochemical course of a reaction ctalyzed by Staphylococcal nuclease. J Am Chem Soc 104:3223.
6. Takahara M, Hibler DW, Barr PJ, Gerlt JA, Inouye M (1985). The ompA signal peptide directed secretion of Staphylococcal nuclease A by _Escherichia_ _coli_. J Biol Chem 260:2670.
7. Hibler DW, Stolowich NJ, Reynolds MA, Gerlt JA, Wilde JA, Bolton, PH (1987). Site directed mutants of Staphylococcal nuclease. Detection and localization by ^{1}H NMR spectroscopy of conformational changes accompanying substitution for glutamic acid 43. Biochemistry 26:6278.
8. Norris K, Norris F, Christiansen L, Fiil N (1983). Efficient site-directed mutagenesis by simultaneous use of two primers. Nucleic Acids Res 11:5103.
9. Kunkel TA (1985). Rapid and efficient site-specific mutagenesis without phenotypic selection. Proc Nat Acad Sci USA 82, 488.
10. Calderon RO, Stolowich NJ, Gerlt JA, Sturtevant JM (1985). Thermal denaturation of Staphylococcal nuclease. Biochemistry 24:6044.
11. Wilde JA, Bolton PH, Dell'Acqua M, Hibler DW, Poumotabbed T, Gerlt JA (1988). Identification of residues involved in a conformational change accompanying substitution for glutamate 43 in Staphylococcal nuclease. Biochemistry 27:4127.
12. Stanczyk SM, Bolton PH, Dell'Acqua M, Pourmotabbed T, Gerlt, JA (1988). Direct observation of multiple environments for the H_δ but not the H_ϵ proton of a histidine residue in Staphylococcal nuclease. J Am Chem Soc 110:7908.
13. Fox RO, Evans PA, Dobson CM (1986). Multiple conformations of a protein demonstrated by magnetization transfer NMR spectroscopy. Nature 320:192.
14. Evans PA, Dobson CM, Kautz RA, Hatfull G, Fox RO (1987). Proline isomerism in Staphylococcal nuclease characterized by NMR and site-directed mutagenesis. Nature 329:266.

Protein and Pharmaceutical
Engineering, pages 35–53

COMBINING SITES AND EPITOPES DEFINED BY MOLECULAR MODELING, PROTEIN ENGINEERING AND NMR

Anthony R.Rees*, Andrew C.R.Martin, Sally Roberts & Janet C.Cheetham

Laboratory of Molecular Biophysics,
University of Oxford,
The Rex Richards Building,
S.Parks Road,
Oxford OX1 3QU,
England

Abstract The structure of the antibody combining site (ACS), either free or complexed with antigen, has been determined by x-ray crystallography for a number of antibodies. However, since x-ray structures accrue at a rather slow pace we have begun to develop algorithms that will allow prediction of the three-dimensional topology of the combining site from a knowledge of the amino acid sequences of the complementarity determining regions (CDR). At the present time, these algorithms incorporate the elements of *ab initio* and "knowledge based" approaches and allow the ACS to be modelled with a fair degree of accuracy. The problems arising in the development of these algorithms will be discussed as will the manner in which protein engineering can be used to test the models produced. In addition, we will present a new and exciting advance in the application of 2-D NMR to the mapping peptide epitopes in solution.

INTRODUCTION

As x-ray structures of Fab fragments and of their antigen complexes become available it should be possible to correlate the observed sequence changes between CDRs of different antibodies with changes in structure. In time, such analyses will lead to an explanation of how CDR sequence determines binding site topography. The fact is, however, that x-ray structures are accruing at a rather slow pace relative to the acquisition of new sequence information. Clearly there is a need for a structure prediction approach until such time as the solving of x-ray structures becomes no more time consuming than cloning a gene and determining its sequence. Such predictive algorithms do exist, but in general are still insufficiently developed to enable combining sites to be generated with any degree of accuracy. Progress is being made in many laboratories, however, and in this paper we shall concentrate on our own contributions indicating the useful role protein engineering can play in testing the predictions of any model.

The antibody combining site is only half the story, however. To model the complex between an antibody and antigen also requires us to know the 3D structure of the particular epitope. This is not trivial. Of course, if the x-ray structure of the complex is known, then the epitope is precisely defined. The norm is one of two alternatives. Either the x-ray structure of the antigen is known in which case the problem is reduced to mapping the antibody combining site onto the antigen. This is also non-trivial, even when serological or other data have identified the approximate epitope. The second situation is where the antigen structure is also unknown. Modelling of both antibody and antigen, required in this instance, is clearly unrealistic. We are addressing both these situations and have developed a method for mapping peptide epitopes by NMR, which should be applicable to any type of antigen that can be mimicked by a fragment of the antigen (e.g. oligopeptide, oligosaccharide or oligonucleotide). In addition, we are developing an algorithm which, given a particular antibody combining site structure, locates the epitope on the surface of an antigen. In its present form this algorithm correctly selects from several thousand different positions of the antigen, the observed orientation of hen egg white lysozyme (HEL) when complexed with each of the antibodies HyHel5 (1), HyHel 10 (2) and D1.3 (3) (Webster,D. et al. in preparation).

MODELLING THE ANTIBODY COMBINING SITE.

Modelling the antibody combining site (ACS) presents a rather specialised problem, compared with other modelling situations. A large proportion of the variable domain, particularly the V_L-V_H interface, is well conserved in sequence and tertiary fold between different antibodies of known structure (4). This so called 'Fv beta-barrel' region (Fig.1) is therefore an ideal candidate for the homology and knowledge-based structure prediction methods (5). The combining site region itself, however, is comprised of six hypervariable loops, the sequence composition of which is, as their name suggests, highly variable. In addition, the degree of conformational flexibility they exhibit, between

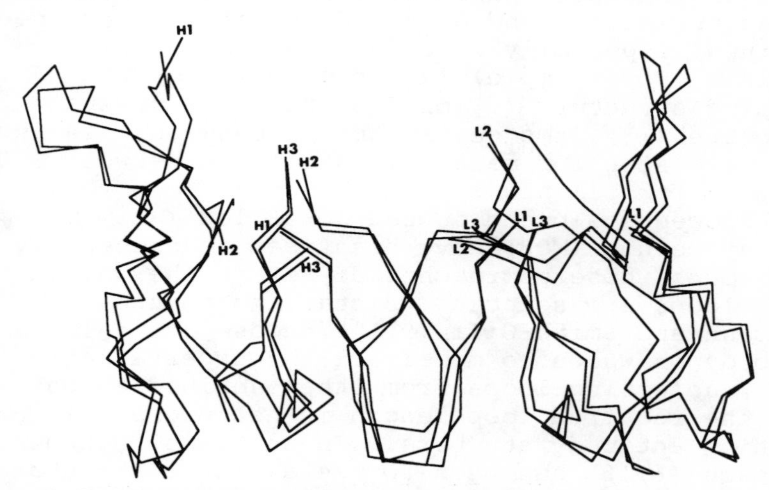

Figure 1. C-alpha superimpositions of Fv framework regions of the antibodies Kol (human) and J539 (mouse) showing the CDR take-off points. The structures are least squares fitted on mainchain atoms by the method of McLachlan (McLachlan,A.D. (1982) Acta Cryst.A38, 871-873).

one antibody structure and another, is large (6). Any modelling procedure for defining the topology of the combining site region must be able to fulfill two conditions. First, to predict accurately the comformations of the individual loops. Second, to reproduce the manner in which the loops pack against one another to generate the antigen binding surface. For a given ACS, a precise knowledge of both mainchain and sidechain conformations is required in order to define the structural origins of the antigen specificity. This is also an important consideration if the modelling is to guide protein engineering studies. By contrast, the accuracy required of a model to be used as a starting structure for x-ray refinement is much lower, since the refinement procedure itself will provide final positions for the sidechains (7). Existing approaches to modelling the antibody combining site are frequently homology based, that is, the available CDRs from known antibody crystal structures are used to generate starting models (6,8-11). So far, with one exception (see later), these methods have achieved only limited success; where crystal structures have become available after modelling has been performed, quite large deviations between predicted and observed structures have been obtained, especially in sidechain conformations (12,13). For example, in a comparison of the modelled structure (9) of J539 with the crystal structure (13), rms deviations of between 1.1A and 4.0A were seen for mainchain atoms and between 2.0A and 6.5A for all atoms.

Procedures based on the so-called maximum overlap method (MOP) have been used independently by two groups whose approaches differ in the way in which loops are selected as starting points. Feldmann and Smith-Gill (9,10) choose a single antibody on which to model all the hypervariable loops, selecting loops from other antibodies only when the required loop length is not present in that starting antibody stucture. In contrast, our own approach (6,16) has been to use a single structure for the framework region only, selected on the basis of high resolution and sequence homology. Individual loops are then selected from a database of all the available antibody crystal structures in the Brookhaven protein databank firstly on length and, if two or more loops of equal length exist, on the basis of sequence homology. These loops are then attached to the framework model. In both methods, if no loop of the required length is available, the loop of closest length to the unknown is used and insertions or deletions are made as required, attempting to retain overall shape and intra-loop hydrogen bonding. Energy minimisation is then applied to the whole structure, either restrained (10) or unrestrained (6), using respectively the CHARMM (14) or GROMOS (15) potentials. For the anti-lysozyme antibody Gloop2 (6) such an approach has been remarkably successful. The model for Gloop2 produced by this MOP method has proved to be extremely close to the crystal structure recently determined in our laboratory (Jeffrey P. et al. unpublished) as shown by a C_a-atom rms deviation of ~1.26A (at the present level of refinement). However, the method relies on a CDR of the appropriate length being present in the antibody database. In Gloop2, those CDRs for which no database structures were available were always short, resulting in a limited number of possible conformations in which the loops could be built. Clearly, this is not always the case.

The more recent modification to these methods devised by Chothia (11,12) also exploits the entire antibody structure database. Loop conformation is predicted on the basis of the presence of critical *key* residues. These are residues which affect loop packing (e.g. bulky residues such as Trp,Tyr or Phe), form hydrogen bonds or salt bridges (e.g. Ser,Thr,Asn,Gln,Asp,Glu), or are able to adopt unusual conformations (e.g. Gly or Pro). In comparing a predicted model for the anti-lysozyme antibody D1.3 with the crystal structure (12), Chothia obtained rms shifts of less than 1.0A for backbone atoms (N,C ,Ca,$C_ß$) except in H1 (2.07A), though predictions of sidechain orientation were poorer. A higher resolution electron density map for D1.3 (the current map is at 2.8A resolution) is required before a more critical evaluation of the accuracy of sidechain prediction can be made. In general, when the *key* residues are present and loops of similar lengths are available in the database, the predictions are relatively good, as indicated above. However, for loops with lengths not represented in the database or lacking the critical residues required for this type of analysis, the predictions tend to be poor. In addition, relatively unconserved residues can have a profound effect on loop packing (16). There are a number of problems with methods which rely solely upon the antibody database, the main one being the limited size of the database itself.

Table 1. Antibody fragment structures presently available in the Brookhaven database. * denotes structures deposited but whose coordinates are not yet available; † denotes C_a-coordinates only.

Brookhaven code	Name	Type	Source	Resolution (Å)
1FB4	KOL	Fab	human	1.9
1FBJ	J539	Fab	mouse	2.6
1MCP	McPC603	Fab	mouse	2.7
2HFL	HyHel-5	Fab	mouse	2.54
3FAB	New	Fab	human	2.0
1MCG	Mcg†	B/J	human	2.3
2RHE	Rhe	B/J	human	1.6
1REI	Rei	B/J	human	2.0
2MCP	McPC603	Fab	mouse	3.1
	D1.3*	Fab	mouse	2.8

When more crystal structures become available, the probability of the database containing a CDR of identical length and high sequence homology to that being modelled and of identifying all *key* residues (as defined by Chothia) will increase. One way of expanding the database from which the CDR's are modelled is to examine loops from all known protein crystal structures, rather than restricting the database to antibodies alone. We have developed an approach to the problem which combines the resource of the complete protein structure database (the knowledge based component) with conformational search algorithms (the *ab initio* component). A general outline of the procedure is given in Fig.2. A processed form of the Brookhaven crystallographic database of protein structures (17) is used to search for a range of 'template' conformations for the unknown CDR. In searching the database, a set of distance constraints must be satisfied by the lower region of the loop, close to the framework. These constraints, which define the general shape of the loop at its take-off region from the framework, have been defined by an analysis of the hypervariable loops of known antibody crystal structures. In this way, for a loop of a particular length, it is typical to extract some 20-30 'template' conformations for an unknown CDR. Each of these loops is then overlapped onto the conserved

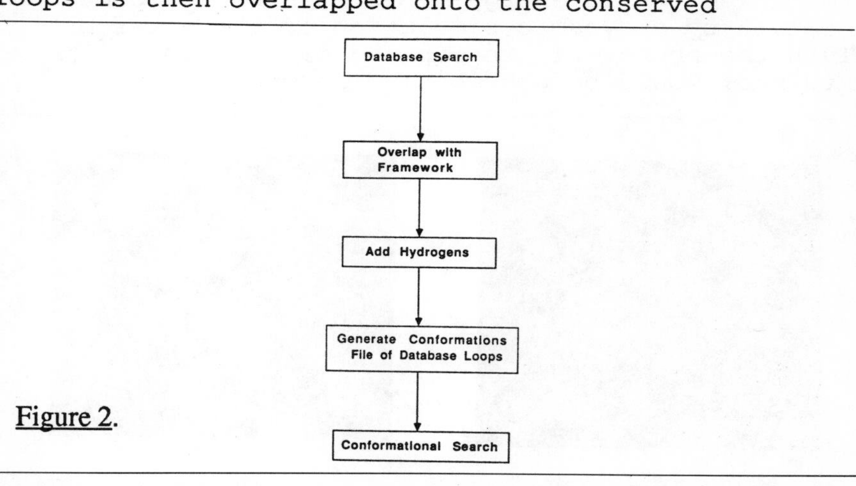

Figure 2.

framework region of the combining site structure and hydrogens are added (using standard hydrogen positions for each amino acid) to provide a set of starting models for the unknown loop confomation. For each model, the residues which are most likely to be involved in antigen contact (generally the tops of the loops) or which are most likely to impart particular conformations on the loop (i.e. glycines and prolines) are removed and reconstructed using the conformational search program CONGEN (18). The resulting conformations can then be ranked in energy by one of three procedures: i) the CHARMM potentials within CONGEN can be employed, though when used *in vacuo* this frequently give a disparity between the lowest energy conformation (Emin) and the minimum RMSD conformation (Rmin). Fig.3a shows the Rmin conformation for H2 from HyHEL 5 which ranks 202nd in energy while Fig.3b shows the Emin structure ranking 366th in RMSD; ii) use of the GROMOS or CHARMM potentials in the presence of a stochastic water box. Although for H2 of HyHEL 5 this had the desired effect of lowering the energy of the Rmin below that of the in vacuo Emin (as in Fig.3b), many other conformations of high RMSD were ranked with even lower energies. This can be explained by the positioning of waters in the stochastic box : some conformations by chance have optimal interactions with the water and as a result are ranked low in energy. This problem could be resolved by a dynamic simulation of the water but, given the number of protein and water atoms involved, would be prohibitive in computer time.

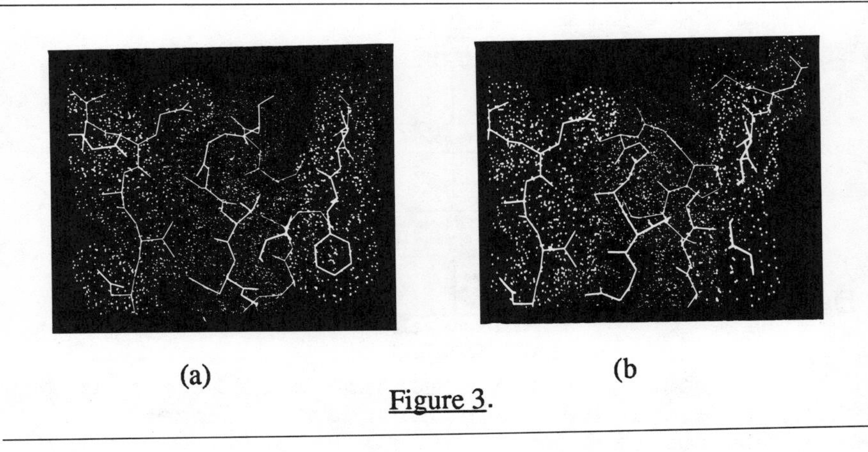

(a) (b

Figure 3.

We have therefore opted for a third procedure; iii) the energy of each conformation is calculated using a substantially modified version of the *in vacuo* GROMOS potentials. At present, we have used this final procedure to model the conformations of all six CDRs in the anti-HEL antibody HyHel 5 with all-atom rms deviations of between 1.5A and 2.5A (Martin et al., in preparation). This result represents a major advance in ACS modelling for two reasons.

First, the procedure is applicable to CDRs of any length and sequence and requires no arbitrary decisions, as are required by the real space renormalisation method presently used by CONGEN in constructing loops of 8 or more residues. Second, one of the CDR conformations showing minimum deviation from the crystal structure is always represented amongst the bottom five energies and can easily be selected on the basis of solvent accessibility. In the above protocol each CDR has been constructed in the presence of the other five, thus taking into account the effects of CDR-CDR interactions during the modelling process. However, when modelling the entire combining site *ab initio*, the order in which the loops are constructed on an 'empty' framework affects the outcome and is a problem that has to be confronted. Our solution to this is to implement a procedure to pack temporary starting models for the loops (defined by database searching) to give an improved model for the environment in which the CONGEN constructions are performed.

In summary, our efforts in Oxford are directed toward producing a general procedure for predicting the 3-D structure of antibody combining sites that uses information from observed loop conformations in all known protein structures and combines it with a sophisticated conformational search algorithm that takes into account not only polypeptide, but also solvent parameters.

THE ROLE OF PROTEIN ENGINEERING

The interaction between an antibody and a protein antigen involves an extensive area of contact (~600A^2) consisting of perhaps a dozen or more side chains contributed by the antigen and extensive use of most or all CDRs by the antibody. Thus, there will be many contacts between the two whose individual free energies of interaction will sum to give the measured free energy of binding, expressed in the equilibrium binding constant. Unfortunately, the individual sidechain (and backbone) contributions to this free energy cannot be determined from simple inspection of the x-ray structure or model, though the overall contributions of favourable hydrophobic and electrostatic terms can be estimated and, using crude analytical methods, allowances can be made for the unfavourable configurational entropy losses incurred on binding (19,20)

One way of attempting to assess particular side chain contributions is to employ site-directed mutagenesis. This approach has been used successfully by Fersht quantitatively to dissect the contributions of substrate binding interactions to catalysis (e.g. 21). In our own studies, we have been forced to employ the technique in a rather more circumspect manner since, until recently, we have been dealing not with a high resolution x-ray structure of the antibody-antigen complex but with a computer model.

The questions we have addressed are :

Is a particular CDR residue, predicted from the model to play a role at the antibody-antigen interface, actually involved?

If so, what is the extent of that residue's contribution to the binding constant?

Though it may seem trivial to arrive at answers to these questions - simply make the mutation and assess the change in binding energy - it is not always obvious in the absence of a structure how to separate effects due simply to changes in surface interactions from those that influence the structure or stability of the antibody combining site itself. Nevertheless, a good deal of information about the probable accuracy of a particular model can be obtained. For example, when assessing the model generated between an anti-lysozyme antibody (Gloop 2) and its epitope (the "loop" region of hen egg lysozyme, HEL) we predicted that electrostatic interactions may play a role in orienting the two molecules in the pre-collision complex (6). To test the prediction two charged residues located at the edge of the combining site (Glu 28 in CDR L1 and Lys56 in CDR H2) were mutated, individually and together. Binding analysis showed that, for the double mutant, the affinity for HEL had *increased 10-fold* and concomitantly, its cross-reactivity with related antigens was reduced (16). Computer modelling of the complex in the presence of these mutations suggested an improved surface of interaction, the total energy of the complex being somewhat lower than the wildtype. Thus, it appeared that the electrostatic interactions proposed were not important for the Gloop 2-HEL interaction.

More recently, mutations of the region of Gloop 2 around position 28 in L1 have suggested that the opposing antigenic surface in this region prefers a hydrophobic environment at the interface. For example, as shown in Figure 4, mutation of Gln 27 (L1) to Leu results in an increase in affinity of 8-fold (an extra 1.22 Kcal/mol) while mutation of Ser 30 to Ala gives a 6-fold increase (1.06 Kcal/mol). Surprisingly, mutation of Glu 28 to Ser gives the same increase in affinity as Glu 28 to Ile (3-4 fold, 0.8 Kcal/mol). Surprising because the side chain of Ile is not only hydrophobic but also larger than Glu and much larger than Ser, and therefore might be expected to cause some disruption at the interface. The very fact that it does not, in spite of potentially losing more configurational entropy than either Glu or Ser on immobilisation at the interface, argues for a relatively loosely packed hydrophobic environment with considerable potential for improvement. However, the most interesting

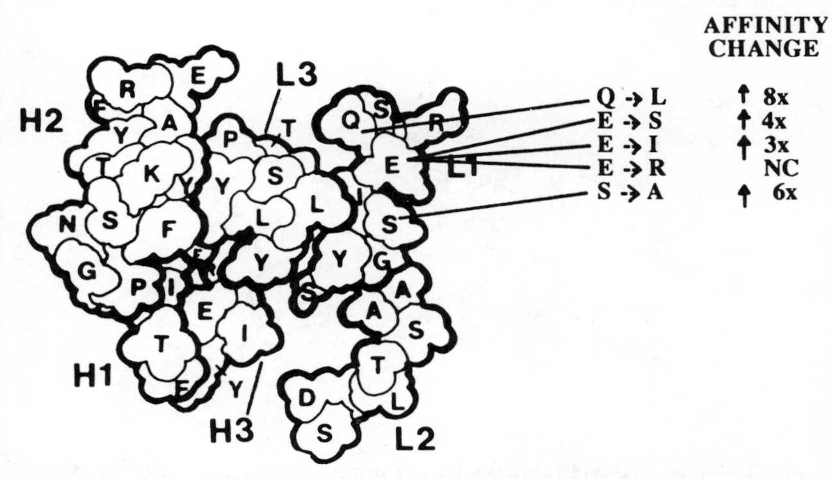

Figure 4. Plan view of the antibody combining site residues of Gloop 2 showing one of the regions of L1 subjected to site-directed mutagenesis, expression and binding analysis.

result was obtained when Glu 28 was mutated to Arg. There are two reasons to suppose that such a mutation would have disrupted the normal interaction. First, there is a charge reversal and second, the Arg side chain is much larger than Glu. In fact, the binding affinity for HEL remained unchanged, the more surprising if indeed the general climate of this region is hydrophobic. The observation could be explained if the arginine side chain was behaving as an amphiphilic residue. Our analysis of arginine side chains interactions at the subunit interfaces of all known structures (Brookhaven database) has indeed shown that this side chain frequently takes part in two types of interaction : 1) the expected electrostatic and/or H-bonding interactions associated with the guanido head and 2) hydrophobic interactions involving the butyl part of the side chain. An example of the latter interaction is shown in Figure 5, taken from aspartate transcarbamylase (22), where a leucine can be seen making a close contact with the gamma and delta CH_2 groups of an arginine in the adjacent subunit. A complete analysis of all such interactions and of those involving lysine side chains will be published elsewhere (Rees & Martin, in preparation). The acceptable substitution of Arg

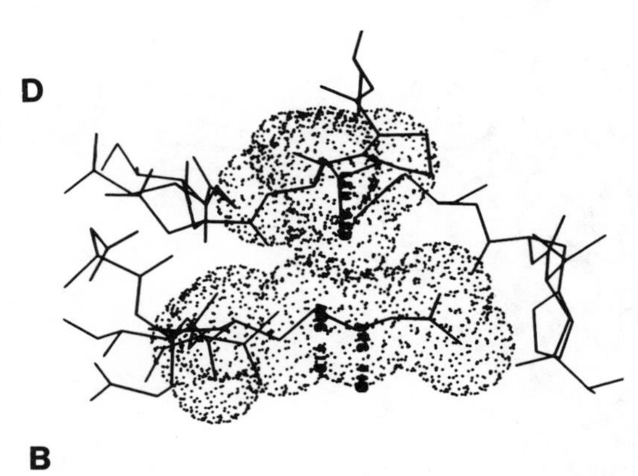

Figure 5. van der Waals surfaces showing the interfacial hydrophobic interactions between the sidechain of leucine 48 in the D subunit and the butyl region of arginine 41 in the B subunit ATCase.

for Glu in the anti-HEL antibody can thus be accommodated by a model wherein the side chain assumes an orientation such that the charged head is allowed access to solvent, or some other hydrophilic shell, while the butyl component sits in a hydrophobic pocket. Whatever the precise orientation, it is clear that a single group hydrophobic potential, such as is assigned to most sidechains when the hydrophobic contribution to free energies is to be estimated, is too simplistic as the simple calculation in Table 2 demonstrates.

The above examples illustrate how careful the antibody engineer has to be in making assumptions about the relative importance of different interactions to an antibody-antigen complex formation. At the same time, a considerable body of information can be obtained by making carefully designed, conservative changes. However, to be certain about whether a particular CDR sidechain makes contact with antigen or not requires some sort of structural analysis. The most obvious answer is to carry out x-ray structural analyses of both native and mutant antibody-antigen complexes. This is time consuming and non-trivial. To reduce the

	ΔG_T (Kcal mol^{-1}) (EtOH -- H_2O; 25^{0}C)
$-CH_2-CH_2-CH_2-NH-C(=NH_2^+)-NH_2$	+ 0.75#
$-CH_2-$	+ 0.73*
$-(CH_2)_3-$	+ 2.19*
$-NH-C(=NH_2^+)-NH_2$	- 1.44*

measured
* calculated

Table 2. Dissection of the arginine side chain in terms of free energies of transfer of the hydrophobic and hydrophilic groups of the same sidechain. The absolute numbers, taken from Cantor & Schimmel (Biophysical Chemistry, 1980, Part I, Freeman) for this particular transfer (EtOH to water) are given merely to illustrate the way in which this sidechain should be considered.

question to the level of particular sidechain involvements we have begun to develop NMR methods that will enable the interactions of sidechains of both antigen and antibody in the complex to be defined.

NMR IN THE STUDY OF ANTIBODY-ANTIGEN INTERACTIONS

Presently, the only direct method of determining an epitope structure is by x-ray crystallography. As previously discussed, this is a time-consuming process and limited in its application to those antibody-antigen complexes for which crystals can be obtained (only four so far). Far more widely used are the indirect methods of a) serological mapping, in which either natural variants of the parent antigen or panels of synthetic peptides are employed; b) chemical modification where accessibility of reactive sidechains (e.g. Lys, Glu or Asp) is determined in the presence and absence of antibody and c) proteolytic digestion experiments in which resistance of surface regions to proteolysis with and without antibody present is used to identify epitope residues. All indirect methods, however, suffer from the disadvantage that they cannot distinguish local effects from those that result from more long range structural modification to the antigen.

To circumvent these problems we have devised a two-dimensional (2-D) NMR method which allows the direct mapping of epitope residues, provided the epitope can be adequately mimicked by a linear peptide. In principle the method could be applied to both continuous and discontinuous epitopes as well as other types of antigenic compounds such as carbohydrate and nucleic acid. An outline of the approach is shown schematically in Figure 6 and can be seen to exploit the difference in mobility between those residues bound to the combining site and those that remain free. The resonances of peptide residues tightly bound to the antibody are broadened in the 2-D NMR spectrum to linewidths comparable to the Fab fragment itself, while unbound residues produce signals comparable with those of

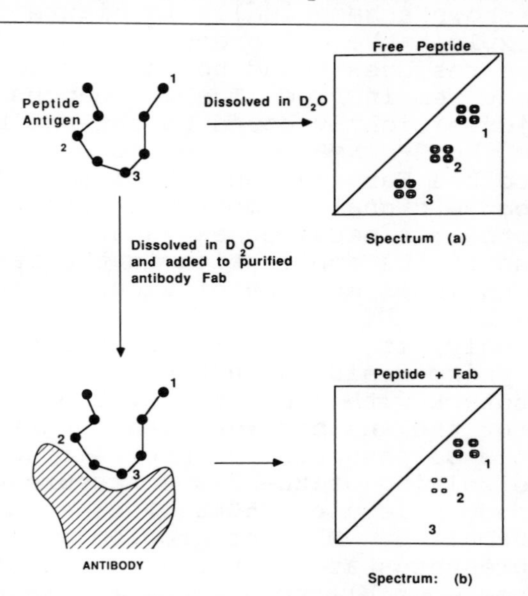

Figure 6. Schematic of a 2D NMR experiment in which the COSY spectra of an antigenic peptide are recorded in the free (a) and bound (b) states. The expected behaviour of resonances in the three different situations are as follows : (1) no broadening beyond that due to slower tumbling of the peptide when in association with the Fab; (2) significant broadening due to partial immobilisation of residues at the edge of the region of contact; (3) substantial broadening due to specific interactions between the ACS residues and those of the peptide comprising the epitope.

both continuous and discontinuous epitopes as well as other types of antigenic compounds such as carbohydrate and nucleic acid. An outline of the approach is shown schematically in Figure 6 and can be seen to exploit the difference in mobility between those residues bound to the combining site and those that remain free. The resonances of peptide residues tightly bound to the antibody are broadened in the 2-D NMR spectrum to linewidths comparable to the Fab fragment itself, while unbound residues produce signals comparable with those of both continuous and discontinuous epitopes as well as other types of antigenic compounds such as carbohydrate and nucleic acid. An outline of the approach is shown schematically in Figure 6 and can be seen to exploit the difference in mobility between those residues bound to the combining site and those that remain free. The resonances of peptide residues tightly bound to the antibody are broadened in the 2-D NMR spectrum to linewidths comparable to the Fab fragment itself, while unbound residues produce signals comparable with those of the free peptide. A detailed explanation of the technique and of its application in the mapping of the HEL discontinous epitope of Gloop 2 can be found in reference 23.

Undoubtedly, if a similar possibility existed for mapping those residues in the antibody that are in direct contact with the antigen, then a complete description of the antibody-antigen interface in solution would be possible. However, if this required the solving of the 3-D structure of an entire antibody molecule (150kDal) or even an Fab fragment (50kDal) by NMR, the problem would be beyond the present state of the art. Fortunately, it has now become possible to express recombinant fragments of antibodies that comprise only the variable domains (Fvs). These fragments, formed by non-covalent association of the light and heavy chain V-domains, have been expressed in E.coli in our laboratory as separate chains followed by reconstitution to active Fvs (24), as dicistronic chains secreted either into the periplasmic compartment (25) or the culture medium (26). Since the individual domains are only about 10kDal each, the prospect of determining the complete structure by NMR is a real possibility. When the complete Fv is assigned, not only will this allow the combining site residues that participate in antigen binding to be identified but it may also allow an investigation of the vexed question of conformational change during complex formation.

Conclusions

We have attempted in this paper to give a picture of some recent published and unpublished advances in molecular modelling and protein engineering of the antibody combining site. While considerable progress has been made in the development of predictive algorithms for the ACS there still remain fundamental difficulties. In particular, good criteria that will allow selection of the correct CDR conformation from a set of energetically close conformers, need to be developed. The problem of CDR-CDR packing is a second question that needs to be addressed. A third question concerns the role of solvent in defining the allowable CDR conformations. Simulation of water surfaces is not straightforward, but clearly the presence of water around and within the ACS and its influence on the dielectric environment of each CDR must be incorporated into future algorithms if a more complete picture of the antibody-antigen interface is to emerge.
The development of a method for mapping epitopes by 2D NMR will make possible, we believe, more rapid advances in molecular modelling of antibody-antigen complexes. It may also become useful in the design of peptide vaccines whose functional role is to mimic the conformation of key regions of the native protein from which they derive.
While site-directed mutagenesis can play a useful role in the analysis of predictions, it is at its most efficient when carried out with reference to three dimensional structures of either antibody, antigen or both. It may be that SDM of antibody and antigen in concert will be the most revealing though experience with proteins whose structures are well defined suggests that interpretation of even conservative substitutions is not always easy.

In conclusion, we confidently predict that in the near future algorithms will emerge that will allow the rational design of antibody combining sites. When this becomes possible, molecular immunology will experience a quiet revolution.

ACKNOWLEDGEMENTS

We would like to thank the SERC (U.K.) for financial support and the MRC for a research studentship to ACRM. We would also like to acknowledge the essential collaboration of Drs.C.Redfield and C.M.Dobson in the NMR studies. We are indebted to Robert Bruccoleri for supplying not only the program CONGEN but also the sourdce code which has allowed us to modify the program to our own needs.

REFERENCES

1. Sheriff,S., Silverton,E.W., Padlan,E.A., Cohen,G.H., Smith-Gill,S.J., Finzel,B.C. & Davies,D.R. (1987) Proc.Natl.Acad.Sci. USA 84, 8075
2. Padlan,E.A., Silverton,E.W., Sheriff,S., Cohen,G.H., Smith-Gill,S & Davies,D.R. Unpublished.
3. Poljak,R.J., Amzel,C.M., Avey,H.P., et al., (1973) Proc.Natl.Acad.Sci. USA 70, 3305
4. Alizari, P., Lascombe, M.B. and Poljak, R. (1988) Ann. Rev. Immunol., 6, 555
5. Sutcliffe,M.J., Haneef,I., Carney,D. & Blundell,T.L. (1987) Protein Eng. 1, 377
6. de la Paz,P., Sutton,B.R., Darsley,M.J. & Rees,A.R. (1986) EMBO J. 5, 415
7. Jensen,L.H. (1985) in "Methods in Enzymology" Vol. pp.227-234.
8. Feldmann,R.J., Potter,M. & Glaudemans,C.P.J. (1981) Molec.Immunol. 18, 683
9. Mainhart,C.R., Potter,M. & Feldmann,R.J. (1984) Molec.Immunol. 21, 469
10. Smith-Gill,S.J., Mainhart,C.R., Lavoie,T.B., Feldmann,R.J., Drohan,W. & Brooks,B.R. (1987) J.Mol.Biol. 194, 713
11. Chothia,C. & Lesk,A.M. (1987) J.Mol.Biol. 196, 901
12. Chothia,C., Lesk,A.M., Levitt,M., Amit,A.G., Mariuzza,R.A., Phillips,S.E.V. & Poljak,R.J. (1986) Science 233, 755

13. Suh,S.W., Bhat,T.N., Navia,M.A., Cohen,G.H., Rao,D.N., Rudikoff,S. & Davies,D.R. (1986) Proteins 1, 74
14. Brooks,B., Bruccoleri,R.E., Olafson,B.D., States,D.J., Swaminathan,S. & Karplus,M. (1983) J.Comput,Chem. 4, 187
15. Aqvist,J., van Gunsteren,W.F., Leifonmark,M. & Tapia,O. (1985) J.Mol.Biol. 183, 461
16. Roberts,S., Cheetham,J.C. & Rees,A.R. (1987) Nature 328, 731
17. Bernstein, F.C. et al. (1977) J. Mol. Biol. 112, 535
18. Bruccoleri,R.E. & Karplus,M. (1987) Biopolymers
19. Janin,J. & Chothia,C. (1978) Biochemistry 17, 2943
20. Wodak,S.J., de Crombrugghe,M. & Janin,J. (1987) Progress in Biophysics & Molecular Biology, , 29
21. Fersht,A.R., Leatherbarrow,R.J. & Wells,T.N.C. (1987) Biochemistry 26, 6030
22. Ke,H-M., Honzatko,R.B. & Lipscomb,W.N. (1984) Proc.Natl.Acad.Sci.USA 81, 4037
23. Cheetham,J.C., Griest,R., Redfield,C. Dobson,C.M. & Rees,A.R., In press.
24. Field,H., Yarranton,G.T. & Rees,A.R. (1988) In "Vaccines '88", Cold Spring Harbor Symposium, pp.29-34.
25. Skerra,A. & Pluckthun,A. (1988) Science 240, 1038
26. Better,M., Chang,P., Robinson,R.R. & Horwitz,A.H. (1988) Science 240, 1041

Protein and Pharmaceutical Engineering, pages 55–70

THE ACETYLCHOLINE RECEPTOR[1]

Michael J. Shuster, Alok K. Mitra,
and Robert M. Stroud

Department of Biochemistry and Biophysics,
University of California, San Francisco
San Francisco, California 94143-0448

ABSTRACT The three-dimensional structure of the nicotinic acetylcholine receptor (AChR) has been determined at 22 Å resolution by combining image reconstruction from tilted views of two-dimensional crystals in the electron microscope with X-ray diffraction data to 12.5 Å from oriented AChR-enriched membranes. Comparisons between structures derived from crystals grown with and without an associated actin-binding 43kD protein reveal the location of this protein which is found to interact both with the cytoplasmic domain of AChR and lipid head groups. Features of AChR structure which help determine ion selectivity and channel conductance have been identified.

INTRODUCTION

A super-family of ion channels is emerging with the discovery that neurotransmitter receptors for acetylcholine, GABA, and glycine, which function as ligand-gated ion channels, have similar sequences, and therefore may share the same functional correlates of structure (1-5). The structure of the nicotinic acetylcholine receptor (AChR) is the closest to being visualized at high resolution, and so we ask how these molecular structures support rapid and selective ion permeation, and how ligand binding is transduced into channel opening. (For recent

---

[1]This work was supported by the National Institutes of Health (GM24485) and by the National Science Foundation (PCM83 16401).

reviews of AChR structure and function, see refs. 1,6,7,8).

The acetylcholine receptor is a 295kD complex of four homologous transmembrane glycoprotein subunits with $\alpha_2\beta\gamma\delta$ stoichiometry (9-11) arranged like staves around a central channel (12,13). Upon agonist binding, the ion channel is opened, allowing the influx of ~10^4 sodium ions/msec/AChR into the cell, depolarizing the postsynaptic membrane (14,15). The subunit arrangement around the channel is unique. The α chains are separated by one other subunit (16-20), and based on immunoelectron microscopy, the most accurately measured angular distance between equivalent epitopes on the two α-subunits is 144° ± 4° (18). This is consistent with a quasi-symmetrical and nearly precise pentameric arrangement (2*360°/5). The subunit arrangement, αβαγδ clockwise as seen from the synapse, was deduced from electron microscopy and chemical crosslinking studies (13,18,20) and is an agreed consensus at this time.

In the mature synapse, the AChR is found associated with several different components, but stoichiometrically so with only one principal cytoskeletal component, a 43kD protein (21,22) that has actin binding activity (23). The associated non-AChR proteins are removed by extracting the membranes at pH 11 for 1 hour (24). There is no evidence that any of these proteins regulate the activity of AChR. The actin-binding 43kD protein is the major component other than AChR observed at sites of innervation on muscle and *Torpedo* electric organ (25,26). Immunoelectron microscopy shows the 43kD protein to be closely associated with AChR (27) and its removal is accompanied by a reduction of material at the cytoplasmic surface that is coextensive with AChR (63). It exists at the synapse at an approximate 1:1 stoichiometry with the AChR (28), has been found to crosslink specifically to the β subunit (29), and interactions between this 43kD protein and AChR seem to encode the relative immobilization of the AChR at the synapse (30,61,62).

Over the past decade in our laboratory and elsewhere, a wide variety of methods have been applied to characterize AChR structure so that we may begin to understand mechanistically its function. High resolution structure determination of AChR requires diffraction quality crystals for X-ray analysis. Significant progress has been made in our laboratory toward attaining this goal. To date, however electron microscopy of two-dimensional crystals combined with X-ray diffraction data from oriented AChR enriched membranes provides the best picture of the three-dimensional shape of the AChR (13,31,38). Diffraction from oriented membranes shows that there are bundles of α-helices within the complex that lie perpendicular to the membrane plane. The ion channel probably is formed within the center of the bundle (32). The central channel is ~7 Å in diameter across the narrow transbilayer region (13), consistent with the sizes of the largest ions that pass through the ion channel (33,34).

Two-dimensional AChR crystals were first found and characterized in native tubular vesicles from *T. californica* (35), and subsequently from *T. marmorata* (36,37) with the associated 43kD protein present. Helical reconstruction of native tubes after ice-embedding, and of already formed tubes dialyzed against pH 11 immediately before freezing was reported (38). However, in those experiments, the samples used for tube formation were alkylated with N-ethylmaleimide prior to alkaline extraction. Under such conditions, the major 43kD protein, while redistributed, still is associated with the AChR enriched membranes (39). To study the structure of the minimal AChR complex, we removed completely the associated proteins from the system prior to tube formation, and then analyses were carried out (31).

METHODS

Image reconstruction of two-dimensional crystals includes information from electron microscopic images of up to ± 52° tilted specimens of crystalline AChR prepared from native and alkali-stripped membranes. Hybrid density maps that include X-ray diffraction data from stacked membranes to 12.5 Å resolution (32,40) were calculated to eliminate some of the distortions introduced in maps based only on limited-tilt angle electron microscopic analyses by including terms in the Fourier summation that correspond to the transform of the electron density perpendicular to the membrane plane. The length of the AChR (perpendicular to the membrane plane) was then seen to be shorter on both the cytoplasmic and extracellular side by 4 to 6 Å each, with respect to the purely electron micrographic reconstruction. Details of these methods are found in reference 31.

RESULTS

Three-Dimensional Structure of AChR.

A section through the middle of the hybrid density map normal to the membrane plane is shown both for native and alkali-stripped samples in Figure 1. The maximum outer diameter is 74 Å for the native, and 70 Å for the stripped AChR. The diameter of the central pore is 24.4 Å for native, and 25.8Å for alkali-stripped AChR (Fig.2). Using the partial specific volume of 0.78 cc/gm (7), the extracellular

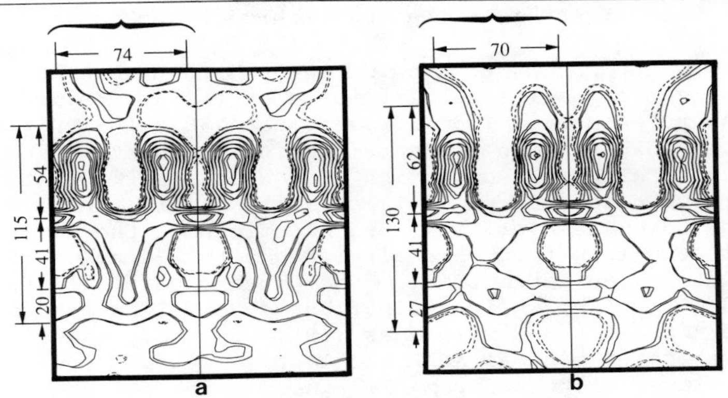

Figure 1. A 4 Å slice normal to the bilayer and through the middle of the hybrid AChR reconstructions both in the native (a) and alkali-stripped AChR (b). The solid lines enclose proteinaceous regions or lipid head groups at the bilayer extremities. The dashed lines above and below the 41 Å bilayer span delineate stain-binding regions, while within the bilayer these represent the low-density terminal regions of the lipids. The dimensions indicated are in Å.

volume of the receptor protein density is ~215,000 Å^3. This corresponds to a molecular weight of 167kD, identical to the predicted molecular weight of the postulated extracellular domain (aligned residue numbers 1-228) of all five glycosylated chains (41,42). The structure of all AChR oligosaccharides have been determined, and two of the putative glycosylation sites on AChR have been investigated by mass spectrometry (43). Consensus sequence number α-143 was found to be glycosylated by a high-mannose oligosaccharide.

The extracellular domain forms a funnel-shaped vestibule (infundibulum), which extends 54 Å above the membrane plane in the native membranes, 62 Å after alkali-stripping (Fig. 1). The dimensions of each subunit above the membrane range from 50 to 60 Å in height. The dimensions of the crossection in projection are 20 Å radial thickness (as if cut like a pie wedge), by 14 Å in contact with the vestibule by 39 Å of outer circumference (Fig. 2), yielding a crossectional area of 575 Å^2 per subunit (above the membrane). Thus, the total contact area between subunits is 1085 Å^2 per interface (vertical height by radial crossection area of the extracellular portion of the density map), or 2170 Å^2 per subunit.

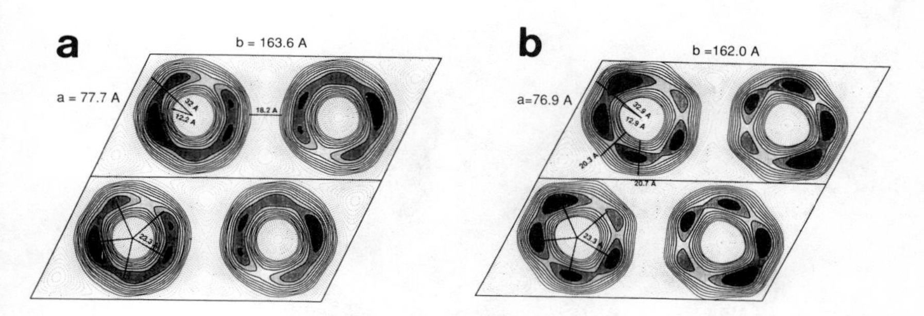

FIGURE 2. Filtered projection images of (a) native and (b) alkali-stripped AChR. The pentameric nature of the structure is apparent in both cases, but more so after removal of peripheral proteins in b. solid lines enclose stain-excluding protein portion of the reconstruction, and the dotted lines indicate the stain-binding, non-proteinaceous regions of the map. The unit cell dimensions for the native AChR are a = 77.7 ± 2.0 Å, b = 163.6 ± 4.0 Å, included angle γ= 114.9 ± 2.5°, and for the alkali-stripped form are a = 76.9 ± 2.0 Å, b = 162.0 ± 3.2 Å, included angle γ= 116.0 ± 1.8°.

The solvent accessible surface area of the total extracellular region is 18,500 Å^2 (3,700 Å^2/subunit), of which ~20%, 3780 Å^2, forms the lining of the infundibulum. The sequences of each AChR chains for different species are more variable by a factor of 2 in the postulated cytoplasmic region, suggesting a less specific role for this domain (1). A solid-view representation of this hybrid density map for a pair of AChR in the unit cell of the two-dimensional crystal is shown in Figure 3.

AChR Subunit Arrangement.

Both native and alkali-stripped forms show the 25 Å wide vestibular entry to the central ion channel as well as a pentameric arrangement of density peaks separated by 72 ± 3° locating the five quasi-symmetric subunits of the AChR (Fig. 2). Columns of density, probably corresponding to the individual subunits, descend slightly inclined to the vertical axis through the AChR, each acting as a nearly

FIGURE 3. A profile outline of one unit cell of AChR from the three-dimensional hybrid density map of the native structure. The surfaces enclose positive densities. The bilayer appears as horizontal planes of densities that correspond to the dense phospholipid head groups that are 41 Å apart. The bundles of cylinders have the dimensions of 38 Å long close-packed α-helices in the center of the bilayer-spanning region, as suggested by X-ray diffraction, and amino acid sequence, and are shown in a cutaway for one of the AChR molecules in the dimer. The figure shows to scale how the closest helices to the channel may lie relative to the entry well. The synaptic side of these helices form the base of the well. The infundibulum extends to just below the surface level of the phospholipid bilayer.

vertical stave around the central channel (13,31,37). The ± 25% variation in extra-membrane stain-excluding volumes of the major peaks around the crest in the projected image (Fig. 2), and the peak heights seen in the three-dimensional reconstruction (Fig. 3), reflect the different sizes of the five subunits, which differ in mass in the extracellular region by 30%, based on their amino acid sequence (1). The five-fold character is more apparent after removal of cytoskeletal and other associated proteins by alkali-stripping (Fig. 2), but dispositions of the major peaks of density, presumed to represent the individual subunits of the AChR, are similar.

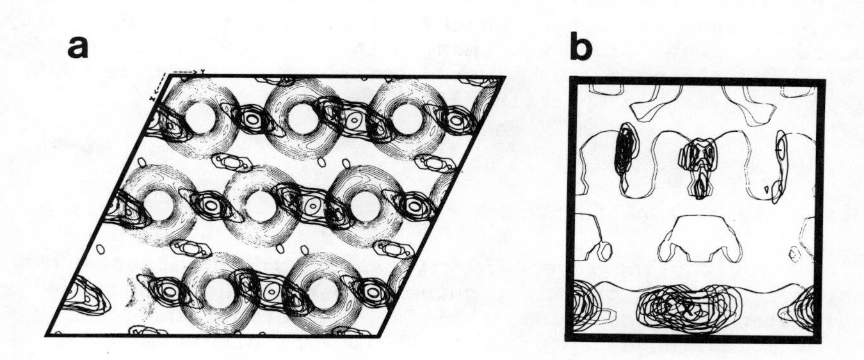

FIGURE 4. Difference stain-exclusion map viewed normal to membrane plane, and looking into the cytoplasmic domain (a), and parallel to the membrane plane (b). (a) is a 50 Å slice of sections 10 Å apart apposed against a lighter image of AChR particles in projection, while (b) is a 20 Å slice along with an outline of AChR particle. The major stain-excluding density marked by X represents the putative cytoplasmic location of the 43kD protein.

Localization of the 43kD Protein.

We find the cytoskeletal 43kD protein to exist in an ordered linkage to AChR. A difference stain-exclusion map (Fig. 4) of native minus alkali-stripped AChR shows that the main density, assigned to the actin-binding 43kD component, is closely associated with the lipid bilayer as well as with the cytoplasmic domain of the AChR. However, it binds beside AChR, not directly under the central channel as suggested by Toyoshima and Unwin (38). This location of the 43kD protein alongside the AChR is more consistent with no known regulatory role for the 43kD protein on AChR channel function (see Discussion).

There is excellent agreement between the volumes of density for structural components and expected volumes based on their molecular weights. After alkali-stripping which removes a significant amount of stain-excluding mass, the volume of stain-excluding material in the cytoplasmic domain lies approximately in a cylinder of radius 33.5 Å, and height of 23 Å for a volume of 81,000 Å^3, equivalent to that expected for 62kD of protein. This agrees with the size of the cytoplasmic domain predicted from models in which each subunit has four (76kD) or five (59kD) α-helical transmembrane crossings. On the cytoplasmic side the density contours always are lower than in the extracellular entry well, indicating a less ordered cytoplasmic structure that is more permeated by stain (31).

Stability of AChR Lattices.

Acetylcholine receptors crystallize in the absence of any cytoskeletal proteins, suggesting that the AChR alone is sufficient to encode and stabilize clustering, and perhaps to do so during synaptogenesis (31). The disulfide bond that crosslinks δ-δ chains of adjacent pentamers in about 80% of native *Torpedo* AChR is not required to stabilize the lattice of AChR; the crystalline tube structures are stable indefinitely.

DISCUSSION

The results of our structural investigations of AChR provide insight into the mechanism of ion channel function. The molecule appears as a quasi five-fold symmetric particle with an overall length of 115 Å. The widest outer diameter of the structure is 74 Å, 38 Å above the lipid phospholipid head groups. After alkali treatment, the extracellular domain increases from 54 Å to 62 Å perpendicular to the membrane, and the entire AChR particle is 130 Å long. The density appears more dispersed at the top and bottom than in the native structure, probably the result of a disordering effect of pH 11.0 treatment, though the in-plane diameter of 72 ± 2 Å is unchanged. A protein wall 24.5 ± 1.5 Å thick surrounds the entry well. This thickness is typical for the dimensions of anti-parallel stranded β-barrel structures as predicted for the extracellular domain of AChR from amphipathic analysis and secondary structure prediction (42,44).

High single-channel conductance, as well as cation selectivity may be a direct functional consequence of the negatively charged vestibule . The calculated pI for the extracellular region (aligned sequences 1-228 of the five subunits containing 150 Glu and Asp, 98 Lys, and 30 His) is

4.77. The net charge on the extracellular domain expected at pH 7.0 is -50. The average charge density is 1 negative, 0.67 positive/25 Å^2 of solvent accessible surface. This represents an almost close-packing of charged residues on the AChR surface. Assuming an even distribution, the expected net charge within the entrance to the AChR channel will be an excess of 10 negative charges. It is likely that the outside of the rosette which contacts the negatively charged lipid head groups carries somewhat more positive charge while the inside surface that forms the entry well is more negative. This excess negative charge density could contribute to the channel's cation selectivity by concentrating cations in the channel mouth. Dani (45), and Dani and Eisenman (46) show quantitatively how negative surface charge within a vestibule about 25 Å by 54 Å can generate many of the observed electrophysiological properties of AChR.

At the level of the membrane surface leading to the central ion channel, the diameter of the entry well reduces from 25 to 18 Å. Given the dimension of the ion channel, there is then a further reduction in pore diameter from 18 to ~7 Å within the transbilayer region. This channel probably is lined by nearly vertical transmembrane α-helices as modeled in Fig. 3. Further support for this model comes from sequence analysis (reviewed in 1) which identifies four hydrophobic stretches of ~25 residues each, which are very likely α-helical, and so of ~40 Å length. This closely matches the measured distance of 41 ± 1 Å between phosphatidyl head groups (47). A total of 20 transbilayer helices (four per subunit) are thought to surround the channel.

Construction of chimaeric AChR's (48), and site-directed mutagenesis (49), has shown that individual negative charges flanking the ends of M2 (the second putative membrane-spanning α-helix) are critical for determining AChR channel conductance, implicating M2 as the helix closest to the channel. Leonard *et al.* (50) find that replacement of serine residues by alanine or phenylalanine at a postulated polar site within M2 also affects channel conductance properties, as well as the binding properties of an open-channel blocking drug. Chemical labeling by molecules thought to interact directly with the ion channel (51-54) also identifies the putative transmembrane helix closest to the channel as M2.

Gating of the AChR may be provided in part by tight ion-binding sites that are located within the transmembrane portion of the resting channel (40). The conformational changes associated with agonist binding are small as assessed by estimates of a net increase in AChR volume of ~80 ± 12 Å^3 accompanying closed to open transitions (55).

Tritium-hydrogen exchange analyses (56) also show only minor differences in the rate of proton exchange from solvent-accessible regions upon exposure to desensitizing concentrations of the agonist carbamylcholine. Only slightly larger changes accompany treatment with activating concentrations. The small conformational changes associated with AChR desensitization have been characterized in three dimensions (57).

Neurotoxins, which lock the channel closed (58) and compete with acetylcholine for high affinity sites on the α-subunits of the receptor are located on the top crest of the synaptic rim (12,18). Thus, the minimum distance over which the signal generated by ligand binding is propagated to open the channel must be at least 54 Å (the length of the vestibule). More significant changes occur in the part of the protein that is in contact with the lipid bilayer and surrounds the ion-conducting channel as assessed by a dramatic decrease in labelling by the lipophilic diazirine TID (3-trifluoromethyl-3-iodophenyl diazirine) (59,60) in the presence of agonists. The observations that conformational differences between resting and desensitized states are small (56), and that small changes in receptor volume accompany closed to open state transitions (55) suggest that a series of subtle conformational changes relay information about ligand binding to the channel gating mechanism. Neurotoxins α-bungarotoxin and curare induce much larger changes in conformation and so must close down channel function by a mechanism different from stabilization of either the resting or desensitized states (56). However, details about these changes as well as the nature of the gate await higher resolution studies of AChR structure.

AChR from *Torpedo* is found *in vivo* primarily as a covalent dimer linked through a disulfide bridge on the δ subunits. The physiological role, if any, of this covalent dimer is unknown. We find that reduction of the dimer to monomer does not affect crystallinity, suggesting that AChR's propensity to close pack may not be the result of covalent dimerization.

Clustering of AChR observed during synaptogenesis may be encoded in AChR itself, as suggested by our observations that formation of tubular lattices is unhindered by the absence of the major associated cytoskeletal 43kD protein. If so, the major role of the actin-binding 43kD protein may be to help anchor these clusters in the postsynaptic membrane.

REFERENCES

1. Stroud R M, and Finer-Moore J (1985). Acetylcholine receptor structure, function and evolution. Annu Rev Cell Biol 1:369-401.
2. Boulter J, Connolly J, Deneris E, Goldman D, Heinemann S, and Patrick J (1987). Functional expression of two neuronal nicotinic acetylcholine receptors from cDNA clones identifies a gene family. Proc Natl Acad Sci USA 84:7763-7767.
3. Whiting P, Esch F, Shimasaki S, and Lindstrom J (1987). Neuronal nicotinic acetylcholine receptor β-subunit is coded for by the cDNA clone α4. FEBS Letters 219:459-463.
4. Schofield PR, Darlison MG, Fujita N, Burt DR, Stephenson FA, Rodriguez H, Rhee LM, Ramachandran J, Reale V, Glencorse TA, Seeburg PH, and Barnard EA (1987). Sequence and functional expression of the $GABA_A$ receptor shows a ligand-gated super-family. Nature (Lond). 328:221-227.
5. Greningloh G, Rienitz A, Schmitt B, Methfessel C, Zensen M, Beyreuther K, Gundelfinger E, and Betz H (1987). The strychnine-binding subunit of the glycine receptor shows homology with nicotinic acetylcholine receptors. Nature (Lond). 328:215-220.
6. McCarthy M P, Earnest JP, Young EF, Choe S, and Stroud RM (1986). The molecular neurobiology of the nicotinic acetylcholine receptor. Annu Rev Neurosci 9:383-413.
7. Popot JL, and Changeux JP (1984). Nicotinic receptor of acetylcholine: Structure of an oligomeric integral membrane protein. Physiol Rev 64:1162-1239.
8. Conti-Tronconi BM, and Raftery MA (1982). The nicotinic cholinergic receptor: Correlation of molecular structure with functional properties. Ann Rev Biochem 51:491-530.
9. Reynolds JA, and Karlin A (1978). Molecular weight in detergent solution of acetylcholine receptor from *Torpedo californica*. Biochem 17:2035-2038.
10. Raftery MA, Hunkapiller MW, Strader CD, and Hood LE (1980). Acetylcholine receptor complex of homologous subunits. Science 208:1454-1457.
11. Lindstrom JE, Merlie J, and, Yogeeswaran G (1979). Biochemical properties of acetylcholine receptor subunits from *Torpedo californica*. Biochem 18:4465-4470.
12. Klymkowsky MW, and Stroud RM (1979). Immunospecific identification and three-dimensional structure of a membrane-bound acetylcholine receptor from *Torpedo californica*. J Mol Biol 128:319-334.

13. Kistler J, Stroud RM, Klymkowsky MW, Lalancette RA, and Fairclough RH (1982). Structure and function of an acetylcholine receptor. Biophys J 37:371-383.
14. Anderson CR, and Stevens CF (1973). Voltage-clamp analysis of acetylcholine produced end-plate current fluctuations at frog neuromuscular junctions. J Physiol (Lond). 235:655-691.
15. Neher E, and Sakman B (1976). Single-channel currents recorded from membrane of denervated frog muscle fibres. Nature (Lond). 260:779-802.
16. Holtzman E, Wise D, Wall J, and Karlin A (1982). Electron microscopy of complexes of isolated acetylcholine receptor,biotinyl-toxin, and avidin. Proc Natl Acad Sci USA 79:310-314.
17. Zingsheim HP, Barrantes FJ, Hanicke W, and Neugebauer D-Ch (1982). Direct structural localization of two-toxin recognition sites on an acetylcholine receptor protein. Nature(Lond). 299:81-84.
18. Fairclough RH, Finer-Moore J, Love RA, Kristofferson D, Desmeules PJ, and Stroud RM (1983). Subunit organization and structure of an acetylcholine receptor. Cold Spring Harbor Symp Quant Biol 48:9-20.
19. Bon F, Lebrun E, Gomel J, van Rapenbusch R, Cartaud J, Popot JL, and Changeux JP (1984). Image analysis of the heavy form of the acetylcholine receptor from *Torpedo marmorata*. J Mol Biol 176:205-237.
20. Kubalek E, Ralston S, Lindstrom J, and Unwin N (1987). Location of subunits within the acetylcholine receptor by electron image analysis of tubular crystals from Torpedo marmorata. J Cell Biol 105:9-18.
21. Cartaud J, Sobel A, Rousselet A, Devaux PF, and Changeux JP (1981). Consequences of alkaline treatment for the ultra-structure of the acetylcholine receptor-rich membranes from *Torpedo marmorata* electric organ. J Cell Biol 90:418-426.
22. Cartaud J, Oswald R, Clement G, and Changeux JP (1982). Evidence for a skeleton in acetylcholine receptor-rich membranes from *Torpedo marmorata* electric organ. FEBS Lett 145:250-257.
23. Walker JH, Boustead CM, and Witzemann V (1984). The 43-K protein, ν_1, associated with acetylcholine receptor containing membranes is an actin-binding protein. EMBO (Eur Mol Biol Organ).J 3:2287-2290.
24. Neubig RR, Krodel EK, Boyd ND, and Cohen JB (1979). Acetylcholine and local anesthetic binding to *Torpedo* nicotinic postsynaptic membranes after removal of nonreceptor peptides. Proc Natl Acad Sci USA 76:690-694.

25. Froehner SC, Gulbrandsen V, Hyman C, Jeng AY, Neubig RR, and Cohen JB (1981). Immunofluorescence localization at the mammalian neuromuscular junction of the M_r 43,000 protein of *Torpedo* postsynaptic membranes. Proc Natl Acad Sci USA 78: 5230-5234.
26. Nghiem HO, Cartaud J, Dubreuil C, Kordeli C, Buttin G, and Changeux JP (1983). Production and characterization of a monoclonal antibody directed against the 43,000-Dalton υ_1 polypeptide from *Torpedo marmorata* electric organ. Proc Natl Acad Sci USA 80:6403-6407.
27. Sealock R, Wray BE, and Froehner, SC (1984). Ultrastructural localization of the M_r 43,000 protein and the acetylcholine receptor in *Torpedo* postsynaptic membranes using monoclonal antibodies. J Cell Biol 98:2239-2244.
28. LaRochelle WJ, and Froehner SC (1986). Comparison of the postsynaptic 43kD protein from muscle cells that differ in acetylcholine receptor clustering. J Biol Chem 261:5270-5274.
29. Burden SJ, DePalma RL, and Gottesman G (1983). Cross-linking of proteins in acetylcholine receptor-rich membranes: Association between the β-subunit and the 43kD protein. Cell 35:687-692.
30. Rousselet A, Cartaud J, Devaux P, and Changeux JP (1982). The rotational diffusion of the acetylcholine receptor in *Torpedo marmorata* membrane fragments studied with a spin-labelled α-toxin: Importance of the 43,000 protein(s). EMBO (Eur Mol Biol Organ). J 1:439-445.
31. Mitra AK, McCarthy, MP, and Stroud RM (1989). Three-dimensional structure of the nicotinic acetylcholine receptor and location of the major associated 43kD cytoskeletal protein, determined at 22 Å by low dose electron microscopy and X-ray diffraction to 12.5Å. J Cell Biol 109:755-774.
32. Ross MJ, Klymkowsky MW, Agard DA, and Stroud RM (1977). Structural studies of a membrane-bound acetylcholine receptor from *Torpedo californica*. J Mol Biol 116:635-659.
33. Huang L-YM, Catterall WA,and Ehrenstein G (1978). Selectivity of cations and nonelectrolytes for acetylcholine activated channels in cultured muscle cells. J Gen Physiol 71:397-410.
34. Dwyer TM, Adams DJ, and Hille B (1980). The permeability of the endplate channel to organic cations in frog muscle. J Gen Physiol 75:469-492.
35. Kistler J, and Stroud RM (1981). Crystalline arrays of membrane-bound acetylcholine receptors. Proc Natl Acad Sci USA 78:3678-3682.

36. Brisson A, and Unwin PNT (1984). Tubular crystals of acetylcholine receptors. J Cell Biol 99:1202-1211.
37. Brisson A, and Unwin PNT (1985). Quaternary structure of the acetylcholine receptor. Nature (Lond). 315:474-477.
38. Toyoshima C and Unwin N (1988). Ion channel of acetylcholine receptor reconstructed from images of postsynaptic membranes. Nature (Lond). 336:247-250.
39. Barrantes FJ (1982). Oligomeric forms of the membrane-bound acetylcholine receptor disclosed upon extraction of the M_r 43,000 nonreceptor peptide. J Cell Biol 92:60-68.
40. Fairclough RH, Miake-Lye RC, Stroud RM, Hodgson KO, and Doniach S (1986). Location of terbium binding sites on acetylcholine receptor-enriched membranes. J Mol Biol 189: 673-680.
41. Noda M, Takahashi H, Tanabe T, Toyosato M, Kikiotani S, Furutani Y, Hirose T, Takashima H, Inayama S, Miyata T and Numa S (1983). Structural homology of *Torpedo californica* subunits. Nature 302:528-532.
42. Finer-Moore J, and Stroud RM (1984). Amphipathic analysis and possible formation of the ionic channel in an acetylcholine receptor. Proc Natl Acad Sci USA 81:155-159.
43. Poulter L, Earnest JP, Stroud RM, and Burlingame A (1989). Structure, oligosaccharide structures, and posttranslationally modified sites of the nicotinic acetylcholine receptor. Proc Nat Acad Sci USA 86:6645-6649.
44. Finer-Moore J, Bazan F, Rubin J, and Stroud RM (1989). Identification of membrane proteins and soluble protein secondary structural elements, domain structure, and packing arrangements by Fourier-transform amphipathic analysis. In Fasman G (ed): "Prediction of Protein Structure and the Principles of Protein Conformation," New York: Plenum Press, in press.
45. Dani JA (1986). Ion-channel entrances influence permeation. Net charge, size, shape, and binding considerations. Biophys J 49: 607-618.
46. Dani JA, and Eisenman G (1987). Monovalent and divalent cation permeation in acetylcholine receptor channels. Ion transport related to structure. J Gen Physiol 89:959-983.
47. Stroud RM, and Agard DA (1979). Structure determination of asymmetric membrane profiles using an iterative Fourier method. Biophys J 25:495-512.
48. Imoto K, Methfessel C, Sakmann B, Mishina M, Mori Y, Konno T, Fukuda K, Kurasaki M, Bujo H, Fujita Y, and Numa S (1986). Location of a δ-subunit region

determining ion transport through the acetylcholine receptor channel. Nature (Lond). 324:670-674.
49. Imoto K, Busch C, Sakmann B, Mishina M, Konno T, Nakai J, Bujo H, Mori Y, Fukuda K, and Numa S (1988). Rings of negatively charged amino acids determine the acetylcholine receptor channel conductance. Nature (Lond). 335:645-648.
50. Leonard RJ, Labarca CG, Charnet P, Davidson N, and Lester HA (1988). Evidence that the M2 membrane-spanning region lines the ion channel pore of the nicotinic receptor. Science 242:1578-1581.
51. Hucho F, Oberthur W, and Lottspeich F (1986). The ion channel of the nicotinic acetylcholine receptor is formed by the homologous helices M II of the receptor subunits. FEBS Lett 205: No 1, 137-142.
52. Giraudat J, Dennis M, Heidmann T, Chang J-Y, and Changeux JP (1986). Structure of the high affinity binding site for noncompetitive blockers of the acetylcholine receptor: serine 262 of the δ-subunit is labeled by [^{3}H] chlorpromazine. Proc Nat Acad Sci USA 83:2719-2723.
53. Giraudat J, Dennis M, Heidemann T, Haumont P-Y, Lederer F, and Changeux JP (1987). Structure of the high-affinity binding site for noncompetitive blockers of the acetylcholine receptor: [^{3}H] chlorpromazine labels homologous residues in the β and δ chains. Biochemistry 26:2410-2418.
54. Oberthur W, Muhn P, Baumann H, Lottspeich F, Wittman-Liebold B, and Hucho F (1986). The reaction site of a non-competitive antagonist in the δ-subunit of the nicotinic acetylcholine receptor. EMBO Jour 5, No 8:1815-1819.
55. Heinemann SH, Stuhmer W, and Conti F (1987). Single acetylcholine receptor channel currents recorded at high hydrostatic pressures. Proc Natl Acad Sci USA 84:3229-3233.
56. McCarthy MP, and Stroud RM (1989a). Conformational states of the nicotinic acetylcholine receptor from *Torpedo californica* induced by the binding of agonists, antagonists and local anesthetics: Equilibrium measurement using tritium-hydrogen exchange. Biochemistry 28:40-48.
57. Unwin N, Toyoshima C, and Kubalek E (1988). Arrangement of the acetylcholine receptor subunits in the resting and desensitized states, determined by cryoelectron microscopy of crystallized *Torpedo* postsynaptic membranes. J Cell Biol 107:1123-1138.

58. Jackson, MB (1984). Spontaneous openings of the acetylcholine receptor channel. Proc Natl Acad Sci USA 81:3901-3904.
59. White BH, and Cohen JB (1988). Photolabeling of membrane-bound *Torpedo* nicotinic acetylcholine receptor with the hydrophobic probe 3-Trifluoromethyl-3-(m-[^{125}I]iodophenyl diazirine. Biochemistry 27:8741-8751.
60. McCarthy MP, and Stroud RM (1989b). Changes in conformation upon agonist binding, and nonequivalent labeling, of the membrane spanning regions of the nicotinic acetylcholine receptor subunits. J Biol Chem 264:10911-10916.
61. Lo MMS, Garland PB, Lamprecht J, and Barnard EA (1980). Rotational mobility of the membrane-bound acetylcholine receptor of *Torpedo* electric organ measured by phosphorescence depolarization. FEBS Lett 111:407-412.
62. Barrantes FJ, Neugebauer D-Ch, and Zingsheim HP (1980). Peptide extraction by alkaline treatment is accompanied by rearrangement of the membrane-bound acetylcholine receptor from *Torpedo marmorata*. FEBS Lett 112:73-78.
63. Sealock R (1982). Cytoplasmic surface structure in postsynaptic membranes from electric tissue visualized by tannic-acid-mediated negative contrasting. J Cell Biol 92:514-522.

Protein and Pharmaceutical Engineering, pages 71–88

HIV1 PROTEASE: BACTERIAL EXPRESSION, PURIFICATION AND CHARACTERIZATION[1]

Lilia M. Babé,* Sergio Pichuantes,* Philip J. Barr,# Ian C. Bathurst,# Frank R. Masiarz# and Charles S. Craik*

*University of California, Departments of Pharmaceutical Chemistry and Biochemistry and Biophysics, San Francisco, CA 94143, #Chiron Corporation, Emeryville, CA 94608

ABSTRACT The protease of human immunodeficiency virus-1 (HIV1) has been expressed as a fusion protein with human superoxide dismutase (hSOD). The 36 kD fusion protein contains the protease-specific cleavage sites corresponding to the gag/protease and the protease/reverse transcriptase junctions. Efficient autoprocessing is observed *in vivo* when expressed in *E. coli.* The 10 kD mature protease is completely soluble and has been purified to apparent homogeneity. Amino and carboxyl-terminal sequencing confirm that correct processing occurs at the expected Phe-Pro bonds to release the 99 amino acid mature enzyme. A 94 amino acid protein was also detected which appears to be the product of autolysis at the Leu5-Trp6 peptide bond of the protease. The purified protease processed *in vitro* a recombinant myristylated gag polypeptide purified from *S.cerevisiae*. The resulting structural proteins released from the polypeptide (nucleocapsid, capsid and matrix) were of the expected molecular weight. The capsid protein was shown to have the expected N-terminus resulting from cleavage at the Tyr138-Pro139 bond. Two synthetic decapeptides, mimicking natural cleavage sites, were used in an *in vitro* assay for rapid and sensitive measurement of proteolytic activity.

[1] This work was supported by NIH grant GM 39552, and by Chiron Corporation. L M B was a recipient of a fellowship from the University of California Task Force on AIDS (F885SF122).

INTRODUCTION

The human immunodeficiency virus (HIV), a member of the retrovirus family, is the etiological agent of the acquired immunodeficiency syndrome (AIDS) (1). As for other retroviruses, initial HIV genome translation yields large precursor polyproteins *gag, gag/pol* and *env* which are proteolytically processed to generate the capsid proteins, the viral enzymes (protease , reverse transcriptase and integrase) and the envelope proteins that are found in mature virions (2). The processing of the *gag* and *gag/pol* polyproteins has been attributed to the virally encoded protease (3), while the processing of the envelope proteins is mediated by a cellular protease (4). Mutations within the protease coding regions of retroviruses, including HIV1 (5), have lead to the formation of noninfectious virions composed of non-processed polyproteins.

Analysis of amino acid sequences suggested that the HIV protease was a member of the aspartyl protease family (6). This suggestion was supported by studies of the inhibitory effect of pepstatin A (7,8) as well as the elimination of activity by mutations within the putative active site of the viral protease (5,8). The recently elucidated X-ray crystal structure of the HIV1 protease has confirmed that this viral protease is a dimeric structure analogous to the 2-domain structure of the pepsin-like proteases (10). Higher resolution data will be required to determine the basis of substrate specificity and catalysis. This information will facilitate the design of specific inhibitors that may abolish enzymatic activity and thereby prevent the polyprotein processing essential for viral maturation. The protease, obtained by chemical synthesis (11,12) or by heterologous expression in bacteria (13,17), has been assayed *in vitro* and *in vivo* for the ability to cleave various viral polypeptide precursors and synthetic peptides encoding putative cleavage sites.

In order to carry out detailed biochemical and structural analyses of the protease, reagent quantities are required. For this purpose, we have expressed the protease as a fusion protein in bacteria. The active, mature protease can be purified to apparent homogeneity, and activity can been measured *in vitro* using a recombinant gag polyprotein as well as synthetic peptides. Chemical analyses have confirmed that processing occurs at the appropriate sites flanking the protease, as well as at the MA/CA (2) junction of the $Pr53^{gag}$ substrate and in a synthetic peptide corresponding to the PR/RT (2) junction. The development of efficient methods for the expression of this viral protease as well as for measurement of its catalytic activity will expedite the structural and functional analysis of the protease and its substrates.

EXPERIMENTAL PROCEDURES

Abbreviations.

DTT (dithiothreitol), EDTA (ethylenediamine tetraacetic acid), HPLC (high pressure liquid chromatography), hSOD (human superoxide dismutase), IPTG (isopropyl beta-d-thiogalactopyranoside), PAGE (polyacrylamide gel electrophoresis), PBS (phosphate buffered saline), PEG (polyethylene glycol), PMSF (phenylmethylsulphonyl fluoride), SDS (sodium dodecyl sulfate), TCA (trichloroacetic acid), TFA (trifluoroacetic acid).

Materials.

Chemicals were purchased from Sigma Co. (St. Louis, MO) unless otherwise specified.

Antibodies.

Rabbit polyclonal antibodies were raised against concentrated culture supernatant from yeast cells expressing and secreting an HIV1 polypeptide encompassing the 78 C-terminal amino acids of the protease and the 37 N-terminal residues of the reverse transcriptase (18) [service provided by BAbCO (Berkeley)].

Plasmid constructions.

Plasmid pSOD/PR179 was constructed for expression of the HIV1 protease in bacteria. The 522 bp BglII/ BalI DNA restriction fragment of the *pol* gene of the ARV-2 SF2 strain (19) was ligated to the NcoI/SalI-digested vector pSODCF2 (20) using synthetic adapters. The synthetic oligonucleotides added to the 5' end of the HIV1 DNA fragment regenerated an NcoI restriction site, encoded a Met, an Ala and three residues of the pol gene product (PheArgGlu), and regenerated the BglII restriction site. The synthetic adapter added at the 3' end of the viral DNA fragment regenerated a BalI restriction site, encoded an additional Pro residue of the *pol* gene and a termination codon and regenerated the SalI restriction site. The resulting plasmid pSOD/PR179 encodes hSOD (153 residues) fused to amino acids 5 (Asp) to 179 (Trp) of the pol region which includes the putative viral protease (21). The carboxyl-terminal Asn and the termination codon of hSOD were replaced by a Met and Ala.

Growth of cells and preparation of DNA-free protein extracts.

E. coli strain D1210 (22) harboring plasmid pSOD/PR179 were grown overnight at 37°C in Luria broth (23) containing ampicillin at 100 ug/ml. Ten liters of overnight cultures were used to inoculate 200 liters of M9 minimal medium (23) and cells were grown for 2 hrs to reach an OD_{600} of 0.4-0.6 . The cultures were then induced with IPTG at 200 uM final concentration and grown for another 4 hrs. The cells were harvested in a continuous flow centrifuge and the pellets were frozen and stored at -20°C.

For a standard preparation, 300 grams of wet cell paste were resuspended in 800 ml of sonication buffer (50 mM Tris-HCl, pH 8.0, 5 mM EDTA, 1 mM DTT, 1 mM PMSF, 100 mM KCl and 0.5% Triton X-100). All operations were carried out at 4°C. Cells were sonicated using a Braun homogenizer equipped with a standard probe. The lysate was centrifuged at 12,000 rpm for 20 min in a Sorvall GSA rotor. Protamine sulfate, dissolved in water and neutralized with NaOH, was added to a final concentration of 0.5% (w/v). After gentle stirring for 15 min, the precipitate was removed by centrifugation at 12,000 rpm for 20 min in a Sorvall GSA rotor. Saturated ammonium sulfate (Schwarz/Mann) was added slowly to the supernatant over a 2 h period with gentle stirring to a final concentration of 40% (w/v). The precipitated proteins were collected by centrifugation at 9,000 rpm in a Sorvall GS3 rotor for 90 min. The protein pellet was resuspended in 100 ml of 20 mM sodium phosphate pH 8.0 and dialyzed exhaustively against 12L of the same buffer.

Cation exchange chromatography.

A CM-Sepharose fast-flow (Pharmacia) column (25cm x 2.6cm) was equilibrated with CM buffer (20 mM sodium phosphate, pH 8.0, 1 mM EDTA, 1 mM DTT, 1 mM PMSF and 0.1% Triton X-100). The dialyzed protein sample was adjusted to the same conditions as the CM buffer and loaded onto the column. After washing the column with 600 ml of CM buffer to remove unbound material, the bound proteins were eluted with a 600 ml linear gradient from 0 to 400 mM NaCl in CM buffer. In order to localize the protease, proteins from the collected fractions were separated on polyacrylamide gels and stained with Coomassie blue or immunoblotted. Fractions eluting between 60 and 110 mM NaCl contained the protease. These were pooled and dialyzed overnight against 10 mM Tris-HCl pH 7.5 buffer.

Preparative isoelectric focusing.

The dialyzed sample was concentrated to 55 ml by placing the dialysis bag containing the sample on a bed of PEG 6000 (Mallinkrodt). The sample was then adjusted to a final ampholyte concentration of 1% pH 3-10 and 0.25% pH 8-10 (BioRad) and loaded on a Rotofor (BioRad) unit. The focusing was carried out according to the manufacturer's specifications and the fractions collected were assayed for protein content, pH and the presence of immunoreactive protease. The fractions in the pH range of 9 to 11, containing most of the protease, were pooled and concentrated against PEG as previously described.

Reverse phase HPLC chromatography of protease.

The concentrated protein sample was loaded onto a reverse phase C_3 column (Zorbax Protein Plus, 4.6 mm x 25 cm, Dupont). The material was fractionated over 60 min (flow rate: 1 ml/min) using a gradient of aqueous solutions of acetonitrile from 25 to 85% containing 0.1% TFA and monitored at 280 nm. The fractions containing HIV protease were verified by PAGE and silver staining (see below) and were lyophilized to dryness and stored at -20°C.

Polyacrylamide gel electrophoresis and protein blotting.

Gel electrophoresis was performed on either 15 or 17.5% SDS-polyacrylamide discontinuous gels (24). In most cases, protein samples were first precipitated by the addition of 1/10th volume of 100% (w/v) TCA containing 5 mg/ml sodium deoxycholate. After incubating for 1h at 0°C, the sample was centrifuged and the pellet was washed with cold acetone, re-centrifuged and dried in a Savant Speed-Vac. The proteins were redissolved in 10 mM Tris-HCl pH 7.5, sample buffer (24) was added and, after heating for 5 min at 95°C, the samples were loaded on SDS-polyacrylamide gels. Proteins were visualized on acrylamide gels by either Coomassie blue or silver staining of the acrylamide gels. For immunoblotting, separated proteins were transferred electrophoretically to nitrocellulose paper (0.45 um pore size, Schleicher and Schuell) in buffer containing 20% methanol, 150 mM glycine and 20 mM Tris-HCl, pH 8.3. The nitrocellulose sheets (blots) were incubated in PBS containing 0.1% Triton X-100 and 1% dry milk (Carnation) for 30 min. The blots were then incubated in antibody-containing sera [either rabbit polyclonal sera against the protease (diluted 1:500) or pooled sera patients (diluted 1:200)] diluted in blot buffer for 30 min. After three separate 5 min washes in PBS with 0.1% Triton X-100, the blots were incubated with horseradish peroxidase anti-IgG antibodies [either goat anti-rabbit (Boehringer Mannheim) or goat anti-human (TAGO)] for 30 min. Following three separate 5 min washes, color was developed by incubating the blots in a solution of hydrogen peroxide and 4-chloro-naphthol in methanol. All the reactions were performed at 37°C. For reference, a prestained marker mix, containing proteins ranging from 3 to 43 kD (Gibco-BRL), was run in parallel.

Protein assay.

Quantitation of the protein content of each sample during the purification was accomplished using the Micro BCA assay kit (Pierce), following the manufacturer's recommendations.

In vitro protease assays.

A myristylated Pr53*gag* polyprotein expressed in yeast and purified to homogeneity (25) was used as substrate for the detection of protease activity in purified or crude protein extracts. The substrate (1 ug) and protease-containing fractions were incubated in 10 mM Tris-HCl, pH 7.0 containing 130 mM NaCl, 1 mM EDTA and 1 mM PMSF at ambient temperature for 24 hours. The reaction products were analyzed by SDS-PAGE in 15% polyacrylamide gels and immunoblotting. Pooled human sera were used to detect the viral proteins and proteins from HIV-infected cells (26) served as markers of viral protein migration. Pepstatin A was added in some assays to inhibit the viral protease. The catalytic activity of purified protease samples towards synthetic peptides was measured using Peptide I (Ala-Thr-Leu-Asn-Phe-Pro-Ile-Ser-Pro-Trp) and Peptide C (Arg-Ser-Leu-Asn-Tyr-Pro-Gln-Ser-Lys-Trp). Aliquots of peptide substrates (10 ug) were incubated with purified protease (0.5 ug) at 37°C for 1 min to 24 h in 50 mM sodium phosphate buffer pH 5.5 containing 5 mM EDTA, 20 mM DTT and 25 mM NaCl in a final volume of 20 ul. The reaction mixtures were fractionated by reverse phase HPLC on a Vydac C_{18} column (250 x 4.6 mm) using an aqueous gradient of acetonitrile from 0 to 100% containing 0.1% TFA and a flow rate of 1 ml/min. The peptides were detected by UV absorbance at 215 nm. The amount of peptide substrate was calculated from the absorption profiles and plotted versus time. A specific activity was calculated from the slope of the curve in the linear range.

Amino acid analysis.

The determination of the amino acid composition of the purified protease was accomplished using the PICO-TAG (Millipore) method. Following acid hydrolysis, the samples were derivatized using phenylisothiocyanate and the phenylthiocarbamyl amino acids were analyzed by HPLC (27).

Amino-terminal sequence analysis.

Amino acid sequences were determined by automated Edman degradation using an Applied BioSystems 470A gas phase sequencer equipped with an on-line 120A phenylthiohydantoin amino acid analyzer. The derivatized amino acids were separated by reverse phase chromatography on a Brownlee C_{18} column according to Hunkapiller (28).

Carboxyl-terminal sequence analysis.

Samples of purified HIV1 protease, prior to or following reduction and carboxylmethylation (29) were digested with carboxypeptidase Y (Boehringer Mannheim). Each sample was dissolved in 0.1 M pyrimidine acetate buffer pH 5.5 containing 3 nM norleucine and 5 ug of carboxypeptidase Y. Aliquots were removed at different time intervals and the reaction stopped by adding formic acid. The samples were then dried *in vacuo* and analyzed for amino acid content as described above.

RESULTS

Expression of HIV1 protease.

The HIV1 protease has been expressed in bacteria using the pBR322-derived expression vector pSOD/PR179 to transform *E. coli* D1210 cells. The viral sequences encoded within the 522 bp BglII to SalI DNA restriction fragment were ligated 3' to the DNA sequences encoding human superoxide dismutase. The expression of this fusion protein was controlled by IPTG induction of the *tac* promoter (Figure 1). Protease expression was monitored by immunoblotting whole cell lysates with antisera generated against concentrated culture supernatants of yeast cells expressing an HIV1 protease construction (Figure 2). Immunoreactive bands of approximately 36 and 10 kD could be detected within 5 to 10 minutes of induction. These polypeptides correspond to the hSOD/protease fusion protein and the mature autoprocessed protease, respectively. The amount of mature protease expressed continues to increase until it reaches a maximum at 4 hours post-induction and is barely detected at 18 hours.

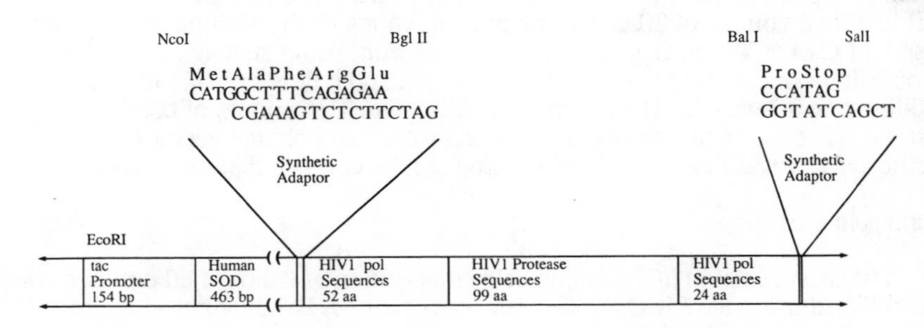

FIGURE 1. Schematic representation of an EcoRI/SalI fragment of recombinant plasmid pSOD/PR179 encoding the HIV1 protease. The restriction fragment encompassing the HIV1 protease was cloned into vector pSODCF2 as described under Experimental Procedures. The beta-lactamase gene used for selection and the ColEI origin of replication used for autonomous replication in *E. coli* are included in the EcoRI/SalI fragment of pSODCF2 that is not illustrated in this figure.

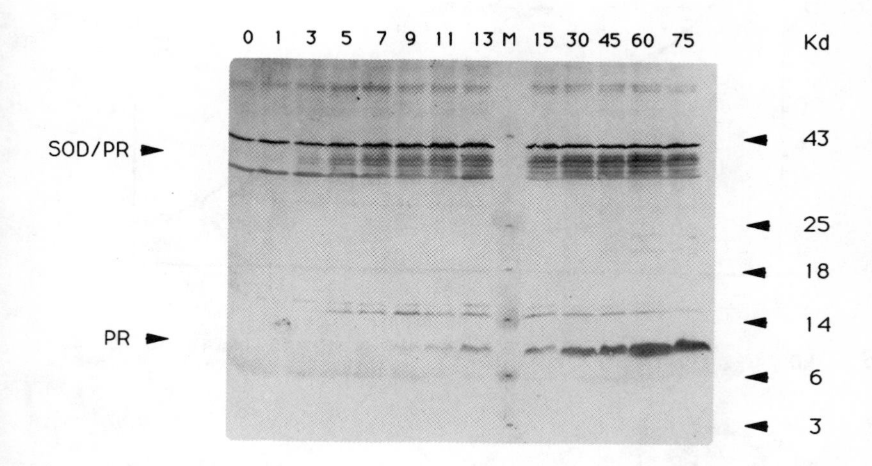

FIGURE 2. Expression of HIV protease in *E. coli* . Cells harboring recombinant plasmid pSOD/PR179 were grown as described under Experimental Procedures. Aliquots were removed at different times post-induction and whole cell extracts were prepared by freezing/boiling of bacterial cells (0.3 OD_{650}) in Laemmli sample buffer. These extracts were separated by SDS-PAGE on 12.5% polyacrylamide gels, transblotted and probed with polyclonal antibodies against the protease.

Purification of the mature protease.

A purification scheme has been developed which results in the isolation of homogeneous HIV1 protease. This scheme, outlined in detail in the experimental procedures section, takes advantage of the strong hydrophobic character and the high isoelectric point of the protease. Efficient extraction of the protease from the bacterial cells by sonication was possible only in the presence of detergent and salt, suggesting strong interactions with the bacterial membranes. No protein denaturants were required during the initial extraction from bacteria, suggesting that the protein was soluble and not in the form of inclusion bodies. The precipitation of the protease at the low ammonium sulfate concentration of 40% was consistent with its predicted hydrophobicity. The calculated isoelectric point (pI) for the protein as determined by its amino acid composition is 9.95, which is markedly higher than the pI values for most bacterial proteins. This property was exploited in the next two purification steps. Most of the proteins flowed through the weak cationic exchange resin, CM-Sepharose, equilibrated at pH 8.0. The protease remained bound and could be eluted at approximately 80 mM NaCl. After desalting the sample, preparative isoelectric focusing in the range of pH 3 to 10 was performed. The majority of bacterial proteins at this stage of the purification display isoelectric points

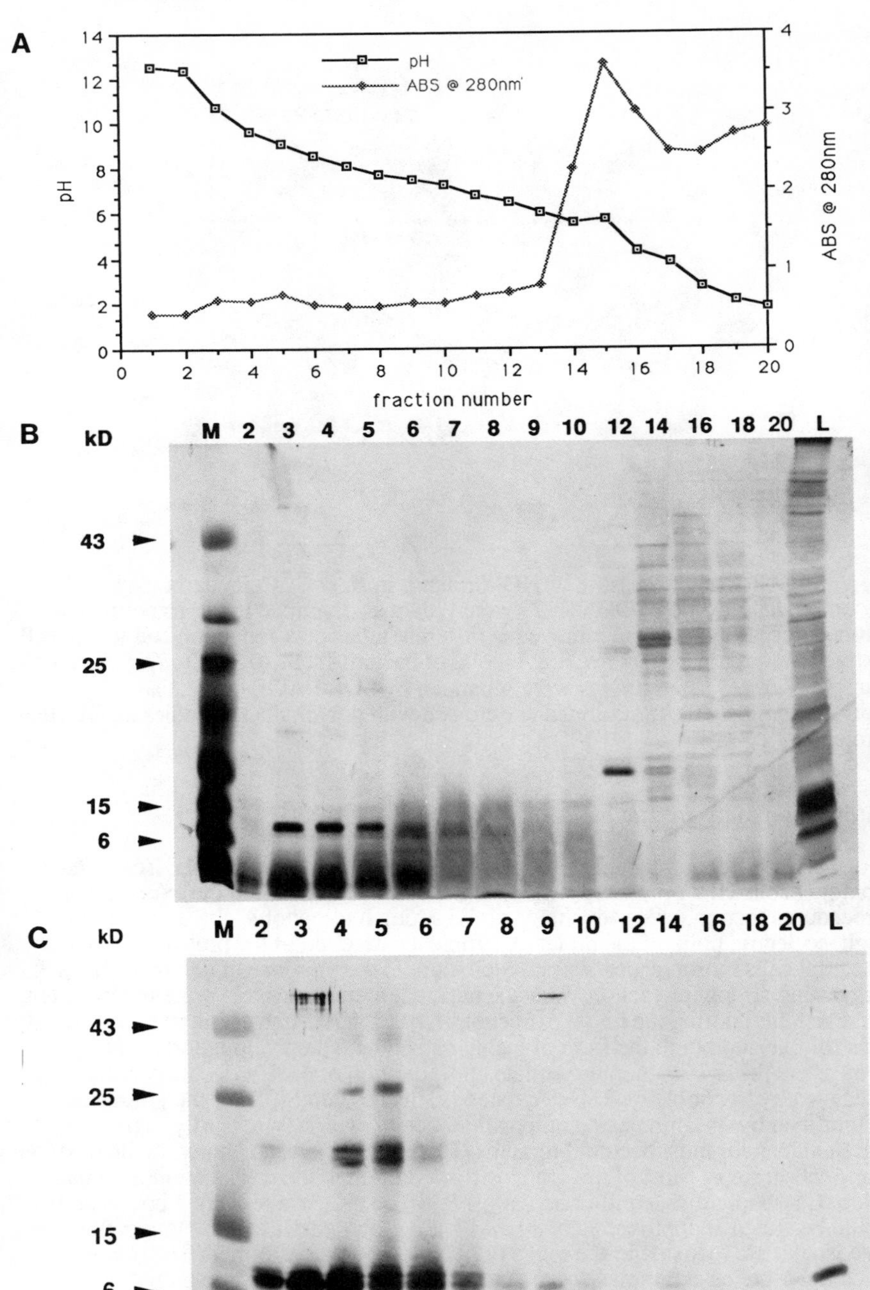

Figure 3. (Legend on page 79)

in the range of 3.5 to 5.0 as demonstrated in Figure 3 panel A, which shows the profile of protein content versus pH for the 20 fractions collected. Aliquots from these fractions were electrophoresed on polyacrylamide gels and stained with Coomassie blue/silver stain (Figure 3, panel B) or immunoblotted and probed with antisera against the protease (Figure 3, panel C). The majority of the HIV1 protease migrated to the region of pH 9 to 11, while the bulk of contaminating proteins migrated to the acidic side, resulting in a 200 fold purification of the protease, based on protein content.The last step in the purification consisted or reverse phase chromatography on a silica column developed with a water/acetonitrile gradient in 0.1% TFA. As expected for a protein of high hydrophobic content, the protein eluted at a high concentration of acetonitrile (Figure 4, fractions 25 through 27). Mass spectrometric analysis of these samples revealed the presence of Triton X-100, suggesting that the protease is purified in detergent micelles. This is an indication of the protein's preference for a hydrophobic environment not unlike the one it would encounter within the membrane of infected cells. Quantitation of the purified protease by the BCA protein assay indicates that approximately 5 mg of protein can be isolated in homogeneous form from about 600 g of wet bacterial cell

Characterization of the purified protease.

The migration of the protease in polyacrylamide gels under denaturing conditions corresponds to a polypeptide of 10 kD as expected for a protein of 99 amino acids. When a sample of the purified protease was subjected to gel filtration in a TSK 250 HPLC column (BioRad), the protein eluted in the region corresponding to a 20 kD polypeptide (data not shown) as expected for the active dimeric form of the enzyme. The amino acid composition of the purified sample as determined by PTH derivatization in a PICO Tag microstation was within 90% agreement with the expected composition. Amino and carboxyl terminal sequencing confirmed that proteolytic autoprocessing occurred at the correct Phe-Pro peptide bonds (Figure 5).

In vitro assays of proteolytic activity.

The catalytic activity of the protease was monitored *in vitro* on the recombinant substrate Pr53^{gag} (25) and on short peptides containing Phe-Pro and Tyr-Pro cleavage sites. The Pr53^{gag} polyprotein used for these assays was expressed in yeast cells, purified and demonstrated to be correctly myristylated (25).

FIGURE 3. Isoelectric focusing of HIV1 protease. The protease-containing fractions isolated from a CM-Sepharose column were subjected to preparative isoelectric focusing on a Rotofor unit (BioRad) using ampholytes of pH range 3 to 10. Panel A - a plot of 280 nm absorbance and pH for the 20 fractions that are isolated from the apparatus (about 2.5 ml each). Panel B - aliquots from each fraction (as the numbers above the lanes indicate) were fractionated by SDS-PAGE on 12.5 % polyacrylamide gels and Coomassie blue/silver stained to detect proteins. Lane M contains pre-stained molecular weight standards and lane L contains an aliquot of the material loaded on the Rotofor unit. Panel C - a gel similar to the one on panel B was immunoblotted with antibodies against the protease.

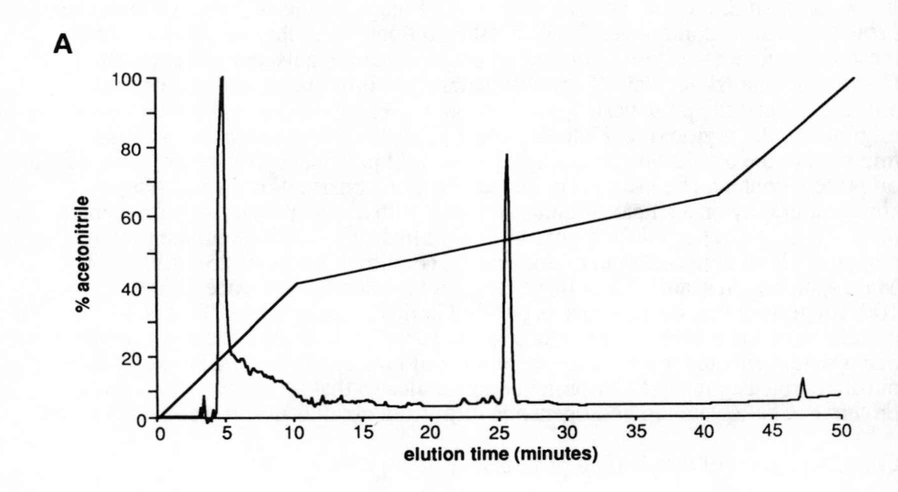

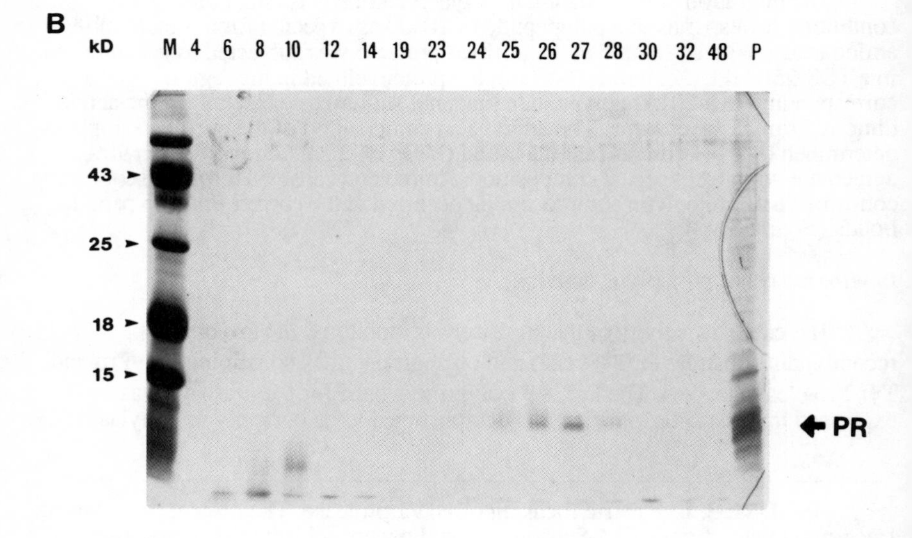

FIGURE 4. Final purification step: reverse phase HPLC. The material isolated by isoelectric focusing was concentrated and loaded onto a C_3 reverse phase column. Protein was eluted at a flow rate of 1 ml/min with a water:acetonitrile gradient containing 0.1% TFA. Panel A shows the eluate absorbance monitored at 280 nm. Aliquots of 1 ml fractions were analyzed by SDS-PAGE on 12.5% polyacrylamide gels using silver stain (panel B).

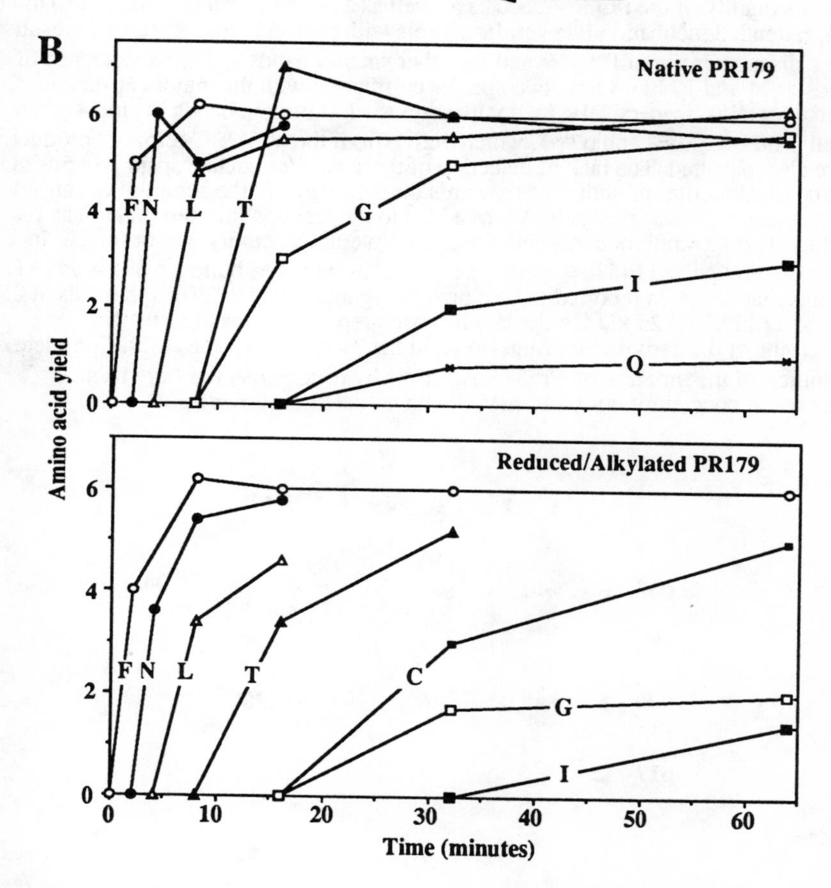

FIGURE 5. Determination of the amino and carboxyl-termini of the purified HIV1 protease. Panel A- The predicted amino acid sequence of the HIV1 protease (21) is presented in the single letter code. The amino acids detected by the sequence analysis are underlined and agree with the predicted amino acid sequence of the *pol* encoded protease. Panel B- Detection of amino acids released by carboxypeptidase Y treatment of purified HIV1 protease (native: top panel, reduced/alkylated: lower panel). Amino acids present in aliquots taken at the times indicated were subjected to amino acid analysis and the yields were normalized to an arbitrary scale of 0 to 7 units.

The *in vitro* assay on purified Pr53^{gag} using purified HIV1 protease was monitored using PAGE of the digestion products and the results of the processing by the protease were observed by immunoblotting using sera from AIDS patients (Fig. 6). The pH optimum for the specific processing was between 6.0 and 7.0 and the sensitivity of this assay was such that as little as 1 ug of Pr53^{gag} was sufficient to monitor specific hydrolysis by the protease. The undigested material, lane 3, consists mainly of the Pr53^{gag} species as well as a ladder of smaller bands (50 to 40 kD). Extended incubation of a similar sample with purified protease (lane 2) results in the disappearence of Pr53^{gag} and the other smaller bands and the appearence of bands at 24 and 17 kD. These two species co-migrate with the major capsid and matrix proteins, respectively, by comparison to viral protein bands in lane 4. Two small proteins, p9^{gag} and p7^{gag}, which derive from the p16^{gag} C-terminal product were also expected. The lack of detection of these smaller nucleocapsid proteins may be due to a low titer of antibodies towards these proteins in the serum of advanced AIDS patients. When pepstatin A was added to the reaction mixture (lane 1) at 1 mM final concentration, partial inhibition of proteolytic activity was observed. In this case, undigested Pr53^{gag} could be detected as well as a build-up of a 40/41 kD species that has been reported to be a processing intermediate (30), and bands at 25, 24 and 17 kD. The 25 kD CA species has been reported to undergo further processing at the carboxyl-terminus to yield the 24 kD CA species (30). Complete inhibition of the protease on Pr53^{gag} substrate by pepstatin A has only been achieved at concentrations of 10 mM or above (data not shown).

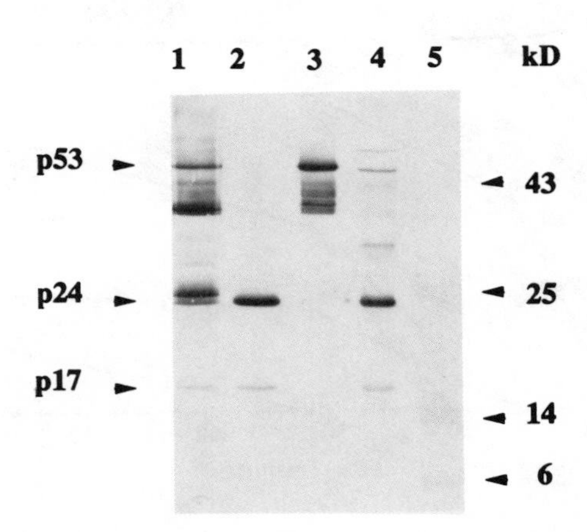

FIGURE 6. Processing of Pr53^{gag} precursor by purified protease. Pr53^{gag} (1ug) was incubated in 10 mM Tris-HCl, pH 7.0, 130 mM NaCl and 1 mM DTT for 24 hours at room temperature in the presence of: 10 ng protease and 1 mM pepstatin A (lane 1), 10 ng protease (lane 2), and no enzyme (lane 3). The digestion mixtures were separated by SDS-PAGE on 12.5% polyacrylamide gels and immunoblotted using sera from AIDS patients. Lane 4: proteins from HIV1 infected cells; lane 5: pre-stained molecular weight markers.

The processing site for the Pr53*gag* at the MA/CA junction was confirmed by N-terminal sequence analysis of the mature capsid protein. The products of a digestion of Pr53*gag* were separated on polyacrylamide gels and transblotted. The bands migrating to 25 and 24 kD were excised and this material was subjected to amino terminal sequence analysis according to the method of LeGendre and Matsudaira (31). The results confirmed the amino termini to be identical for these two peptides and in agreement with the expected hydrolysis of the peptide bond between Tyr138 and Pro139 of the gag precursor. Similar results were recently published for the 25 and 24 kD CA species isolated from virus particles (21, 30).

A faster and more quantitative *in vitro* assay of the proteolytic activity was developed using short synthetic peptides as substrates. Peptide I is a decapeptide that exhibits the sequence found at the processing site between the protease and reverse transcriptase (Ala-Thr-Leu-Asn-Phe-Pro-Ile-Ser-Pro-Trp). Peptide C is a decapeptide that was synthesized to include a "consensus" sequence likely to be hydrolyzed by the HIV protease (Arg-Ser-Leu-Asn-Tyr-Pro-Gln-Ser-Lys-Trp). Figure 7 (panels A and B) illustrates the separation by HPLC of substrates and products from the digestion of synthetic Peptide I with a sample of purified protease. The appearence of the substrate as a doublet, in panel A, may be due to a partial deamidation at Asn4. The products of the digestion, labeled in panel B, were identified by amino acid analysis and confirm that processing occurred as expected at the Phe-Pro peptide bond. Incubation of Peptide I at pH 5.5 with purified protease results in the cleavage of 20 nanomoles of peptide/minute/mg viral protease, while incubation of Peptide C under the same conditions results in cleavage of 54 nanomoles/minute/mg protease.

DISCUSSION

Enzymatically active HIV1 protease expressed in bacteria has been successfully purified to homogeneity. Construction of a fusion protein between hSOD and the protease was required to enhance the expression yield. Analysis of the amino and carboxy termini have confirmed that correct autocatalytic processing of the original translation product occurs in order to yield a 99 amino acid polypeptide. The protease can be extracted in a soluble form from bacterial cells and purified to apparent homogeneity. The yields from the bacterial preparations are approximately 5 mg of purified protease from 600 grams of packed cells.

The purification protocol takes advantage of the uncommonly high isoelectric point and the significant hydrophobicity of the enzyme. The purity of the final product was first assessed by silver staining of polyacrylamide gels and later confirmed by amino acid sequence and composition analyses. The purified material exists within detergent (Triton X-100) micelles that seem to preserve the protease in a stable, active conformation. Disruption of the micelles by detergent removal or replacement severely reduces the enzymatic activity, which can later be regained by the addition of Triton X-100. Extended storage of the protease in the absence of a reducing agent leads to the formation of disulfide bridges. This cross-linked protein migrates as a 20 kD species in non-denaturing polyacrylamide gels and lacks proteolytic activity. Carboxyl-terminal sequence data from carboxypeptidase Y digestion of such a sample failed to detect Cys95, suggesting that it was involved in a disulfide bond (Figure 5, panel B). The proximity of the Cys95 residues in the

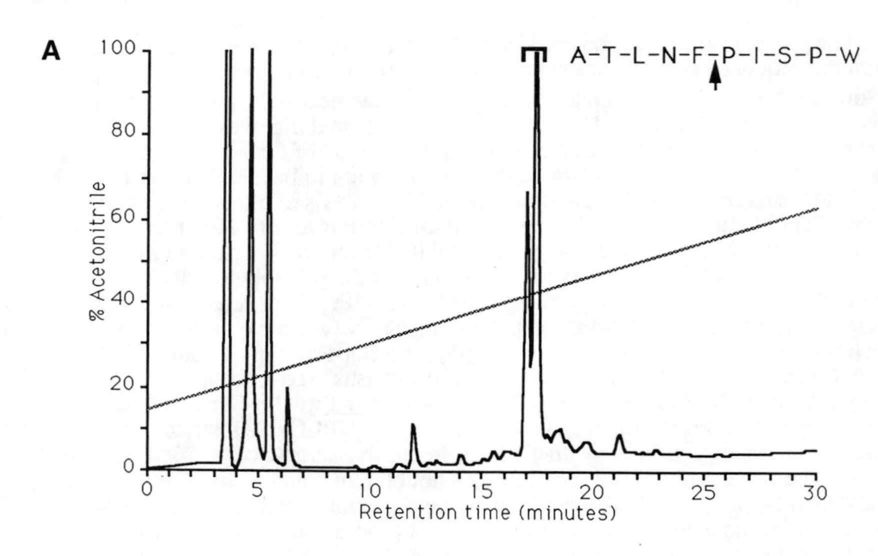

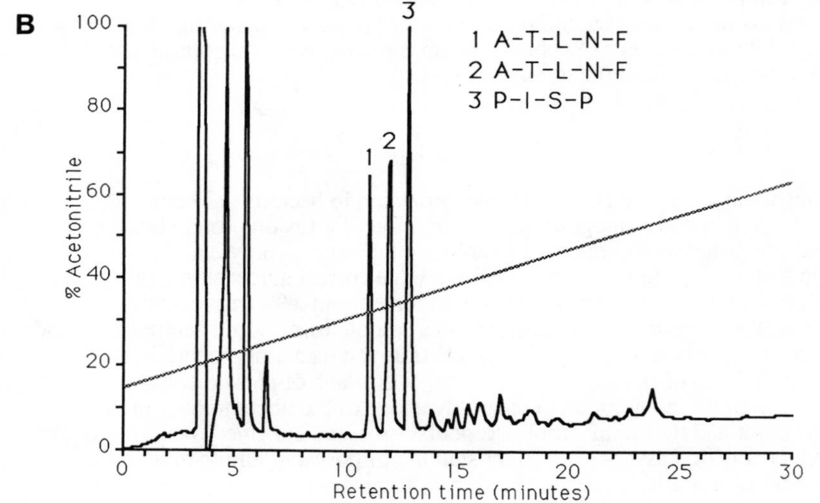

FIGURE 7. Separation by reverse phase HPLC of synthetic peptides following proteolytic digestion. Panel A shows the profile of absorbance monitored at 215 nm for Peptide I in digestion buffer (50 mM sodium phosphate, pH 5.5, 5 mM EDTA, 20 mM DTT, 25 mM NaCl) in the absence of enzyme. The sample was loaded on a Vydac C_{18} column and eluted with a 15-80% linear gradient of water:acetonitrile in 0.1% TFA at 1 ml/min. Panel B depicts the products obtained when 10 ug of Peptide I were incubated in the presence of 0.2 ug of protease for 18 hours at room temperature.

dimeric crystal structure (10), would allow the formation of intermolecular disulfide bonds that may affect catalysis. Addition of a reducing agent (DTT) restores enzymatic activity and allows detection of Cys95 in C-terminal sequence analysis. This reversible loss of enzymatic activity by the protease, exhibited in an oxidizing environment, could be another level of regulation of its function.

The purified protease sample consists mainly of a 99 amino acid peptide, with the expected sequence beginning at Pro1 and ending at Phe99 (Figure 5). A 94 amino acid peptide which is truncated at the amino terminus when cleavage occurs between Leu5 and Trp6 has also been detected (Figure 5). Attempts to separate the 99 and 94 amino acid species by chromatography or differential immunoprecipitation have failed, suggesting that these peptides may exist as heterodimers. The crystal structure of the HIV1 protease as determined by Navia et al. (10) failed to show the location of the initial 5 residues. These are the very 5 residues removed by the internal cleavage we observe suggesting that this very flexible region is accessible to self-processing. Whether these residues are dispensable for structural integrity and proteolytic activity is currently under investigation.

Our *in vitro* assay using $Pr53^{gag}$ indicates that the HIV1 protease is able to correctly process this polyprotein into the components found in mature virions (as determined by immunoblotting as well as N-terminal sequencing of products). We have noticed a progressive order in the processing of $Pr53^{gag}$ during time course digestions (data not shown) which lead to the formation of several peptide intermediates. These same intermediates can be detected in experiments using pepstatin A to decrease enzymatic activity (Figure 6). Polypeptides of approximately 40 to 41kD were first detected which can arise from cleavage at the CA/NC site. Processing to release the N-terminal myristylated MA seems to occur next, followed by trimming of the 25 kD CA species to generate the mature 24 kD CA. The nature of the 24 kD CA has been shown by N-terminal sequence analysis and confirms that the expected cleavage at a Tyr-Pro bond between MA and CA occured. This same order for relative cleavage was observed by Krausslich et al. (8) using synthetic peptides as substrates. The measurement of K_m and k_{cat} for all potential substrates may confirm the role of each substrate sequence in enzyme specificity and the order in which the mature proteins are released.

The other *in vitro* assays we have employed measure the proteolytic cleavage of synthetic peptides by HPLC. This method allows us to quantitfy the turn-over rate for a wide range of peptide substrates under different incubation conditions. Using decapeptides containing Phe-Pro or Tyr-Pro sequences, proteolytic activity was monitored under various conditions of pH, ionic strength and temperature. Although an exhaustive analysis has not yet been conducted, we have determined the optimal pH for peptide digestion to be between 4.0 and 5.5, the salt requirement to be low (between 25 and 100 mM NaCl) and the temperature to be 37°C. The specific activity is in the range of 20 to 50 nmoles peptide per minute per mg of protease. These values are in the range of those observed by others (8) for their preparations of bacterially expressed HIV protease. The differences in turn-over rates obtained in various laboratories may be due to the different reaction conditions employed, the mode of enzyme purification and the viral strain origin of the protease. These discrepancies will be clarified once enzyme preparations are accurately quantitated by such methods as active site titrations and when a unified assay procedure is adopted.

The pH optimum observed for digestion of short synthetic peptides was between 4.0 and 5.5, whereas specific processing of the natural substrates $Pr53^{gag}$ and the reverse transcriptase 66 kD species (data not shown) occur preferentially between pH 6.0 and 7.0. At lower pHs, the natural substrates are cleaved at additional sites, suggesting that unfolding due to denaturation exposes new potential targets for proteolysis. The low pH optimun observed *in vitro* with synthetic peptides may not reflect a physiological situation unless processing occurs in an acidified cellular compartment.

Our ability to generate reagent quantities of protease that can be purified to homogeneity in an active form will allow us to carry out detailed biochemical analysis of its catalytic function. The availability of large quantities of gag precursor substrates and synthetic peptide substrates that can be used in *in vitro* assays will also allow the study of substrate specificity and preferential processing by this enzyme.

ACKNOWLEDGMENTS

We thank Dr. Charles Kettner (Dupont, Central Research and Development) for providing synthetic Peptide I and Erin Bradley (UCSF) for providing synthetic Peptide C.LMB would like to thank John Vásquez for computer assistance.

REFERENCES

1. Popovic M, Sarngadharan M, Read E, Gallo R (1984). Detection, isolation, and continuous production of cytopathic retrovirus (HTLV-III) from patients with AIDS and Pre-AIDS. Science 224:497.
2. The nomenclature adopted for the virally encoded enzymes is that suggested by Leis et al. (32): gag polyprotein precursor, $Pr53^{gag}$; matrix protein (MA, p17); capsid protein (CA, p25/24); nucleocapsid protein (NC, p9, N-terminal product from the cleavage of an initial $p16^{gag}$ which originates from the C-terminus of $Pr53^{gag}$); protease, (PR, p10); reverse transcriptase (RT, p66/51); integration protein (IN, p34).
3. Coffin J (1982). in *RNA Tumor Viruses,* eds. Weiss R, Teich N, Varmus H, Coffin J: Cold Spring Harbor Laboratory, Cold Spring Harbor, NY, p 261.
4. McCune J, Rabin L, Feinberg M, Lieberman M, Kosek J, Reyes G, Weissman I (1988). Endoproteolytic cleavage of gp160 is required for the activation of human immunodeficiency virus. Cell 53:55.
5. Kohl N, Emini E, Schleif W, Davis L, Heimbach J, Dixon R, Scolnick E, Sigal I (1988). Active human immunodeficiency virus protease is required for viral infectivity. Proc Natl Acad Sci USA 85: 4686-4690.
6. Pearl L, Taylor W (1987). A structural model for the retroviral proteases. Nature 329:351.
7. Seelmeier S, Schmidt H, Turk V, von der Helm K (1988). Human immunodeficiency virus has an aspartic-type protease that can be inhibited by pepstatin A. Proc Natl Acad Sci USA 85:6612.
8. Krausslich H-G, Ingraham R, Skoog M, Wimmer E, Pallai, P, Carter C (1989). Activity of purified biosynthetic proteinase of human immonudeficiency virus on natural substrates and synthetic peptides. Proc Natl Acad Sci USA 89:807.

9. Le Grice SF, Mills J, Mous J (1988). Active site mutagenesis of the AIDS virus protease and its alleviation by trans complementation. EMBO J 7:2547.
10. Navia M, Fitzgerald P, McKeever B, Leu C-T, Heinbach J, Harber W, Sigal I, Darke P, Springer J (1989). Three-dimensional structure of aspartyl protease from the human immunodeficiency virus HIV-1. Nature 337:615.
11. Schneider J, Kent S (1988) Enzymatic activity of a synthetic 99 residue protein corresponding to the putative HIV-1 protease. Cell 54:363.
12 Nutt R, Brady S, Darke P, Ciccarone T, Colton D, Nutt E, Rodkey J, Bennett C, Waxman L, Sigal I, Anderson P, Veber D (1988). Chemical synthesis and enzymatic activity of a 99-residue peptide with a sequence proposed for the human immunodeficiency virus protease. Proc Natl Acad Sci USA 85:7129.
13. Debouck C, Gorniak JG, Strickler JE, Meek TD, Metcalf BW, Rosenberg M (1987). Human immunodeficiency virus protease expressed in *Escherichia coli* exhibits autoprocessing and specific maturation of the gag precursor. Proc Natl Acad Sci USA 84:8903.
14. Graves MC, Lim JJ, Heimer EP, Kramer RA (1988). An 11-kDa form of human immunodeficiency virus protease expressed in *Escherichia coli* is sufficient for enzymatic activity. Proc Natl Acad Sci USA 85:2449.
15. Giam C-Z, Boros I (1988). *In vivo* and *in vitro* autoprocessing of human immunodeficiency virus protease expressed in *Escherichia. coli.* (1988). J Biol Chem 263:14617.
16. Krausslich H-G, Schneider H, Zybarth G, Carter C, Wimmer E (1988). Processing of in vitro-synthesized *gag* precursor proteins of human immunodeficiency virus (HIV) type 1 by HIV proteinase generated in *Escherichia coli.* J Virol 62:4393.
17. Darke P, Leu C-T, Davis L, Heinbach J, Diehl R, Hill W, Dixon R, Sigal I (1989). Human immunodeficiency virus protease. Bacterial expression and characterization of the purified aspartic protease. J Biol Chem 264:2307.
18. Barr PJ, Power MD, Lee-Ng CT, Gibson HL, Luciw PA (1987). Expression of active human immunodeficiency virus reverse transcriptase in *Saccharomyces cerevisiae*. Bio/Technology 5:486.
19. Luciw PA, Potter SJ, Steimer K, Dina D, Levy JA (1984). Molecular cloning of AIDS-associated retrovirus. Nature 312:760.
20. Steimer KS, Higgins KW, Powers MA, Stephans JC, Gyenes A, George-Nascimento C, Luciw PA, Barr PJ, Hallewell RA,Sanchez-Pescador R (1986). Recombinant polypeptide from the endonuclease region of the acquired immunodeficiency syndrome retrovirus polymerase (*pol*) gene detects serum antibodies in most infected individuals. J Virol 58:9.
21. Sanchez-Pescador R, Power MD, Barr PJ, Steimer KS, Stempien MM, Brown-Shimer S, Gee W, Renard A, Randolph A, Levy JA, Dino D, Luciw PA (1985). Nucleotide sequence and expression of an AIDS-associated retrovirus (ARV-2). Science 227:484.
22. Sadler JR, Tecklenburg M, Betz J (1980). Plasmids containing many tandem copies of a synthetic lactose operator. Gene 8:279.
23. Maniatis T, Fritsch EF, Sambrook J (1982). Molecular cloning. A laboratory manual: Cold Spring Harbor, N.Y.
24. Laemmli U (1970). Cleavage of structure proteins during the assembly of the head of bacteriophage T4. Nature 227:680.

25. Bathurst IC, Chester N, Gibson HL, Dennis AF, Steimer KS, Barr PJ (1989). N-myristylation of the human immunodeficiency virus type-1 gag polyprotein precursor in *Saccharomyces cerevisiae*. J Virol 63:3176.
26. Steimer KS, Puma JP, Power MD, Powers MA, George-Nascimento C, Stephans JC, Levy JA, Sanchez-Pescador R, Luciw PA, Barr PJ, Hallewell RA (1986). Differential antibody responses of individuals infected with AIDS-associated retroviruses surveyed using viral core antigen p25gag expressed in bacteria. Virol 150:283.
27. Bidlingmeyer BA, Cohen SA, Tarvin TL (1984). Rapid analysis of amino acids using pre-column derivatization. Journal of Chromatography 336:93.
28. Hunkapiller MW (1985). Applied Biosystems- User Bulletin Number 14: Applied Biosystems, Foster City, CA.
29. Babé LM, Brew K, Matsuura SE, Scott W. (1988). Epitopes on the major capsid protein of Simian Virus 40. J Biol Chem 264:2665.
30. Mervis R, Ahmad N, Lillehof E, Raum M, Salazar R, Chan H, Venkatesan S (1988). The *gag* gene products of Human immunodeficiency virus type 1: alignment within the *gag* open reading frame, identification of posttranslational modifications, and evidence for alternative *gag* precursors. J Virol 62:3993.
31. LeGendre N, Matsudaira P. (1988). Direct protein microsequencing from immobilon-P transfer membranes. BioTechniques 6:154.
32. Leis J, Baltimore D, Bishop J, Coffin J, Fleissner E, Goff S, Oroszlan S, Robinson H, Skalka A, Temin H, Vogt V (1988). Standardized and simplified nomenclature for proteins common to all retroviruses. J Virol 62:1808.

Protein and Pharmaceutical Engineering, pages 89–103

STEREOSPECIFIC CATALYSIS BY AN ANTIBODY

Andrew D. Napper, Stephen J. Benkovic
and Patricia A. Benkovic

Department of Chemistry, The Pennsylvania
State University
University Park, Pennsylvania 16802

Richard A. Lerner

Department of Molecular Biology, Research
Institute of Scripps Clinic
La Jolla, California 92037

ABSTRACT Monoclonal antibodies were raised to a transition-state analog that is representative of both a six-membered ring cyclization reaction and ring-opening of the cyclized product. One antibody acted as a stereospecific enzyme-like catalyst for the appropriate substrates. Formation of a single enantiomer of a δ-lactone from the corresponding racemic δ-hydroxyester was accelerated by about a factor of 790. Stereospecific bimolecular aminolysis of the lactone was also catalyzed by the same antibody. The observed turnover rate for the two reactions may be approximated from the measured difference between the binding of reactants and the transition state analog.

INTRODUCTION

Antibodies are a class of biological receptors characterized by enormous diversity and specificity. Different permutations of gene segments coding for immunoglobin peptide chains, and of the chains themselves give a minimum of 10^8 different antibody molecules; this number is expanded further by mutation. The specificity of a given antibody for the corresponding antigen is reflected by association constants ranging from 10^4 to 10^{12} M^{-1} (1). Thus the introduction of catalytic activity into antibody binding sites offers the possibility of a vast range of catalysts with enzyme-like specificity.

One strategy used to generate catalytic antibodies is to exploit the binding specificity to selectively stabilize transition state configurations. It was Linus Pauling in 1948 (2) who first described the importance of transition-state stabilization to enzyme catalysis. He observed that the essential difference between antibodies and enzymes is the type of structures to which they bind. Thus, antibodies are usually found to bind to stable, long-lived structures, whereas enzymes bind preferentially to the high-energy, short-lived transition-state of a chemical reaction. This difference reflects the biochemical functions of antibodies and enzymes. Antibodies locate and bind structures foreign to a given organism, whereas enzymes catalyze chemical reactions by stabilizing the transition state and thereby lowering the energy barrier between reactants and products. Following Pauling's observation, Jencks (3) and others have speculated as to whether an antibody binding a high-energy chemical intermediate would be capable of catalyzing a chemical reaction. Practical development of this idea has recently become possible due to advances in two areas: firstly, the ability to elicit the required immune response is dependent on the availability of stable transition-state analogues of chemical reactions; there are now a considerable number of such species (4). Secondly, obtaining antibodies with a single, defined specificity has been possible only since the development by Kohler and Milstein of protocols for the production of monoclonal antibodies.

Recently a number of groups have obtained catalytic antibodies using haptens rationally designed as transition-state analogs. To date this approach has produced antibodies catalyzing acyl-transfer (5-13) and pericyclic reactions (14-16).

RESULTS AND DISCUSSION

In common with other workers in this field, we chose to study acyl-transfer reactions because phosphonates have been shown to be good mimics of the tetrahedral transition-states in these reactions. The system we chose to study enabled us not only to look for acceleration of a simple one-substrate reaction, but also to examine how far antibodies

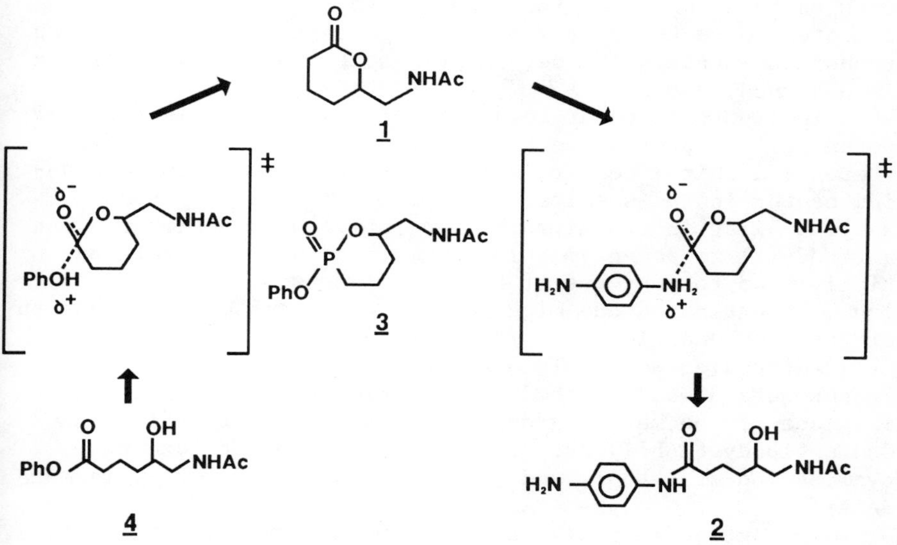

Figure 1. The transition states for the cyclization of ester 4 to lactone 1, and ring opening of the latter by 1,4-phenylenediamine to give amide 2, are mimicked by the cyclic phosphonate, 3. The transition states shown are not necessarily those in the rate-determining step of each reaction.

are able to display two crucial attributes of enzymes. The first of these is stereoselectivity; the ability of an enzyme to selectively bind and catalyze the reaction of one out of a pair of substrates with the same chemical composition, but differing in their three-dimensional structures. The second is the ability of enzymes to sequester two substrates into the same or adjacent binding sites, and thereby facilitate a bimolecular reaction.

Our first step was synthesis of the cyclic phosphonate, 3, which we designed as a transition state analogue for the two reactions shown in Figure 1. The first of these is the cyclization of the hydroxy-ester 4, giving lactone 1 with the concomitant release of phenol. Secondly, we investigated the ring opening of the lactone 1 by aromatic amines, such as 1,4-phenylenediamine, which gives the amide, 2, as shown. Each reaction proceeds through tetrahedral transition-states such as those shown in the Figure. For each reaction it is expected to be the energy of binding to the rate determining transition state which determines whether antibody catalysis is observed.

An important point to notice on this scheme is that the potential for stereoselectivity arises from the fact that there is a chiral center in the hapten, 3, and a corresponding center in the species, 4, 1, and 2. Thus all these species exist in two mirror-image forms, or enantiomers. An antibody, like an enzyme, is asymmetric, and may preferentially bind to just one of the enantiomers of hapten 3. Thus, by extension one might expect catalysis of the reaction of just one enantiomer of substrate 4 or 1.

Antibodies were raised as described previously (9). Phosphonate 3 was attached to a carrier protein via a linker in place of the methyl group in the acetamido function shown. One antibody (24B11) out of 24 was found to catalyze the cyclization of 4 to 1, and we proceeded to study this in more detail.

The antibody catalyzed cyclization may be described by a simple Michaelis-Menten scheme, as follows:

$$\mathrm{Ab} + \mathrm{S} \underset{}{\overset{K_m}{\rightleftharpoons}} \mathrm{Ab.S} \xrightarrow{k_{cat}} \mathrm{Products}$$

Where the velocity of the catalyzed reaction is given by the Michaelis-Menten equation:

$$V = \frac{k_{cat}[\mathrm{Ab}][\mathrm{S}]}{K_m + [\mathrm{S}]}$$

Thus there appears to be a rapid equilibrium binding of substrate and antibody, followed by conversion of bound substrate to product. A characteristic of Michelis-Menten kinetics is that of saturation. The rate of reaction does not increase linearly with substrate concentration but reaches a maximum once all available binding sites are filled. This is apparent from a plot of 1/velocity vs. 1/substrate concentration, the so-called Lineweaver-Burk plot (Figure 2). Thus the straight line obtained is indicative of

Michaelis-Menten kinetics. The intercept on the horizontal axis is the reciprocal of the K_m, which may be taken to be the equilibrium constant for the formation of the antibody-substrate complex.

The intercept on the vertical axis is $1/V_{max}$; the value

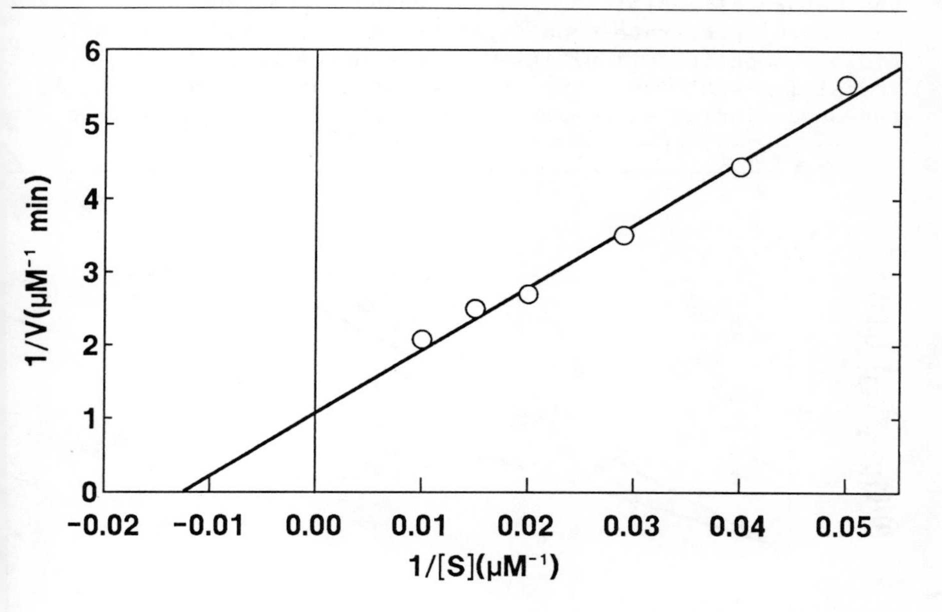

Figure 2. Lineweaver-Burk plot of cyclization of hydroxyester 4 catalyzed by antibody 24B11. Velocities were determined as described previously (10).

of V_{max} divided by the antibody concentration gives k_{cat}, which is the first order rate constant for conversion of bound substrate to products.

An important test of whether the observed catalysis is occurring in the antigen-binding site is provided by studying the inhibition by the transition-state analogue, the cyclic phosphonate hapten. One way of displaying the inhibition data is the Henderson plot, as shown in Figure 3. The y and x variables are linearly related such that data at any substrate concentration should give a straight line intersecting the y-axis at the same point. Furthermore, the value of this intercept is equal to the concentration of binding sites. Thus this gives a method of determining the concentration of catalytically active antibody. The variation of the gradient with substrate concentration indicates the nature of inhibition; an increase in gradient with increasing substrate concentration, as here, is characteristic of competitive inhibition. In other words, the transition-state analogue and the substrate are competing for the same binding site. The inhibition constant K_i, may be

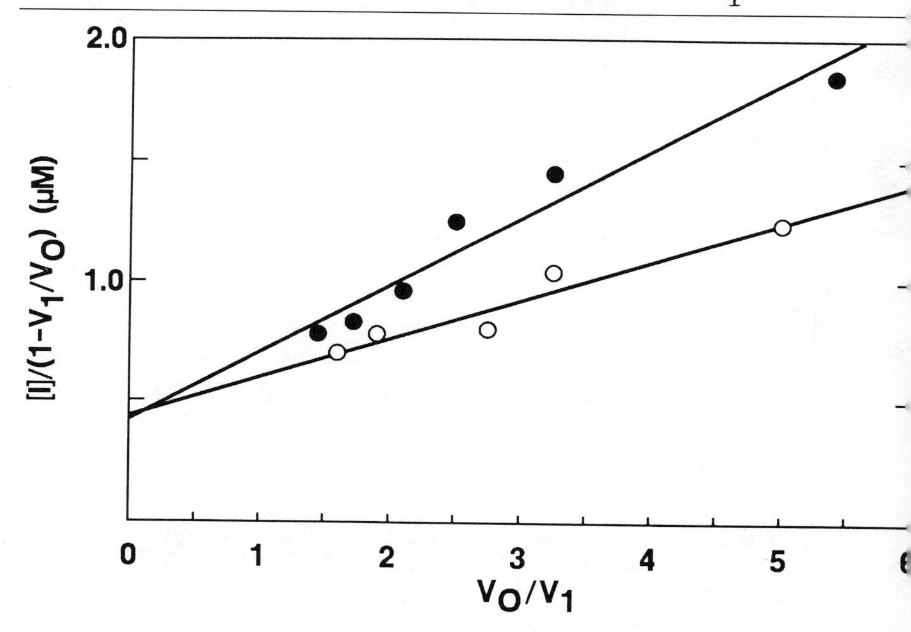

Figure 3. Henderson plot of inhibition of 24B11-catalyzed cyclization of ester 4 (S) by phosphonate 3 (I): •, 200 μM S and 0.25-1.50 μM I; o, 100 μM S and 0.25-1.00 μM I. Velocities were determined as described previously (10).

obtained from the gradient of each line. The value obtained from this graph, together with the other kinetic parameters already discussed, are shown in Table 1. The K_m

TABLE 1.
KINETIC PARAMETERS FOR THE CYCLIZATION OF ESTER 4 BY MONOCLONAL ANTIBODY 24B11.

K_m^* (μM)	76
K_i (nM)	75
V_{max} (μM min^{-1})	0.99
k_{cat} ($minute^{-1}$)	2.36
k_{uncat} ($minute^{-1}$)	0.003
k_{cat}/k_{uncat}	790

*In the absence of binding by the unreactive enantiomer, $K_m \approx 38$ μM for the reactive substrate.

for the substrate is 76 μM, and the k_{cat} for conversion of bound substrate to product is 2.36 min^{-1}. The rate constant for the corresponding background reaction is 0.003 min^{-1}; thus the rate acceleration, k_{cat}/k_{uncat}, is 790. The inhibition constant, K_i, for the transition state analogue inhibitor is determined to be 75 nm. K_i may be taken to be the actual binding constant for the inhibitor; thus the transition-state analogue inhibitor binds 1000 times more tightly than the substrate. This suggests that the antibody does indeed accelerate the reaction by tight-binding, and hence stabilization, of the transition state.

Probably the most exciting observation regarding the cyclization reaction was that the catalysis is indeed enantioselective. In other words the antibody catalyzes the reaction of only one out of the two enantiomers of the substrate. We turned over a relatively large quantity of substrate with antibody, and isolated the product lactone in pure form. The relative amounts of the 2 enantiomers were determined by NMR, using a chiral lanthanide shift reagent. In the absence of any chiral species in solution, the chemical shift of a given proton is the same in both enantiomers. Therefore we used a chiral shift reagent, which coordinates to the amide oxygen. This causes unequal shifts in the resonances of nearby protons in the two enantiomers.

Figure 4 shows a representative section of the NMR of the lactone, showing the peaks corresponding to one of the

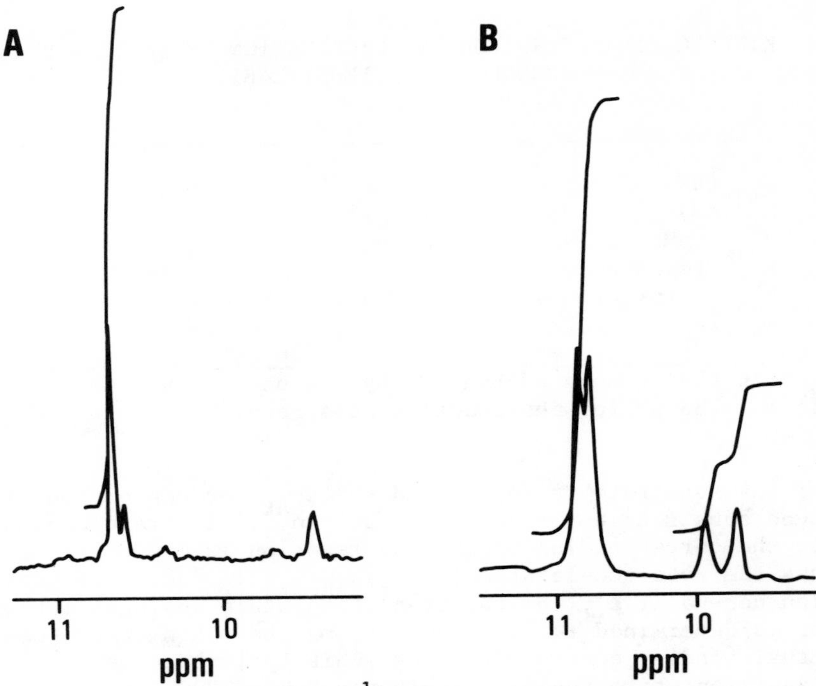

Figure 4. Part of the ^1H NMR (360 MHz, $CDCl_3$) spectrum for the two enantiomers of 1 in the presence of ca. 1 equivalent of tris[heptafluoropropylhydroxymethylene)-d-camphorato] europium (III). Chemical shifts are shown in ppm downfield from TMS. Peak assignments, and the chemical shift differences between enantiomers ($\Delta\Delta\delta$) are as follows: (A) δ 9.45, 9.68 (one of $C\underline{H}_2NHAc$, $\Delta\Delta\delta$ = 0.23), 10.60, and 10.67 ($NHCOC\underline{H}_3$, $\Delta\Delta\delta$ = 0.07); (B) δ 9.71, 9.94 (one of $C\underline{H}_2NHAc$, $\Delta\Delta\delta$ = 0.23), 10.74, and 10.82 ($NHCOC\underline{H}_3$, $\Delta\Delta\delta$ = 0.08. The lactone 1 was obtained as described previously (9).

side-chain methylene protons, and the amide CH_3. Spectrum B shows the NMR of a chemically synthesized, racemic mixture of the 2 enantiomers, hence the pairs of equal size peaks. Spectrum A shows the NMR of a sample of lactone isolated from the antibody-catalyzed reaction; different peak sizes are now clearly visible, for the CH_3 group and the one methylene proton. Integration of the peaks, and subtraction of the amount of both enantiomers expected from the background reaction, indicates that the antibody produces lactone in 94% enantiomeric excess. In other words one of the enantiomers accounts for 97% of the product. This is an important observation, given the extent of current interest in asymmetric catalysis.

We now turned out attention to an energetically more demanding reaction, namely the ring-opening of the lactone by aromatic amines. We chose to study the reaction of 1,4-phenylediamine rather than aniline due to the enhanced reactivity of the former. We were gratified to find that the binding interactions were sufficient for the antibody to overcome the unfavorable entropy change, and sequester two substrates in the same binding site, and catalyze the bimolecular ring-opening as shown in Figure 1.

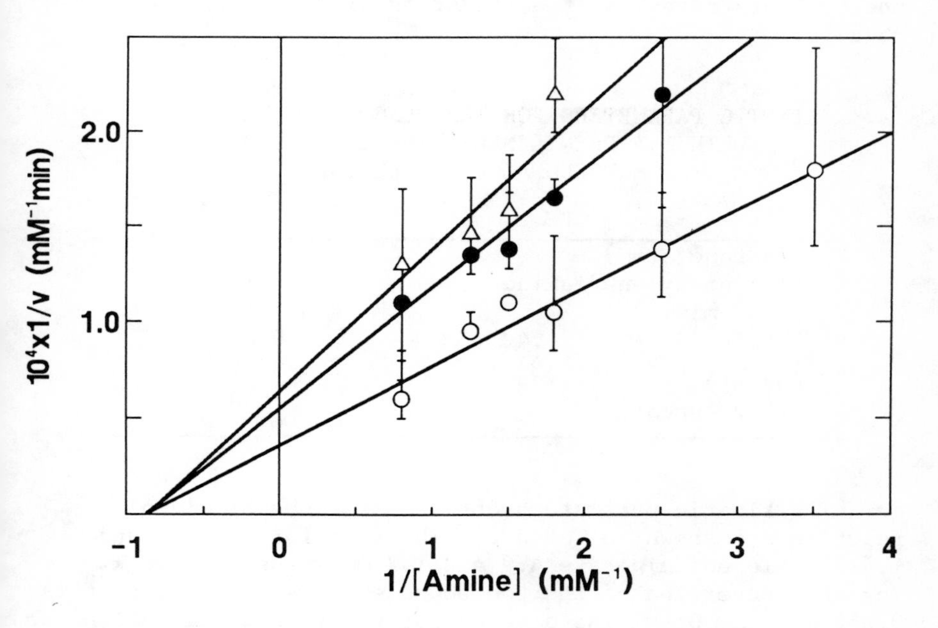

Figure 5. Lineweaver-Burk plot of reaction of 1,4-phenylenediamine (0.28-1.2 mM) with lactone 1, catalyzed by antibody 24B11. o, [lactone] = 15.1 mM; •, [lactone] = 7.5 mM; Δ, [lactone] = 5.0 mM. Velocities were determined as described previously (10).

The catalyzed reaction may be described by simple Michaelis-Menten kinetics (Figure 5), and saturation is observed with respect to both substrates. Since the family of plots in the Figure intersect on the horizontal axis, the binding of one ligand has no effect on the other; thus catalysis of the reaction is not influenced by the order of binding of the substrates. Furthermore, the reaction is strongly inhibited by the transition state analogue. In this case there is also significant product inhibition; the amide product has a K_i ≈35 μM. The substrates were present in the millimolar range; thus the reaction was completely shut down after the 1st 5-10% by product binding to the active site. This may be a general problem with catalysts in which two substrates react to form a single product; entropy loss on binding the single product is much less than that of binding the two substrates, so the former is likely to bind more readily.

TABLE 2.
KINETIC PARAMETERS FOR THE REACTION OF LACTONE 1 WITH 1,4-PHENYLENEDIAMINE CATALYZED BY MONOCLONAL ANTIBODY 24B11.

K_m (lactone) (mM)	4.9
K_m (1,4-phenylenediamine) (mM)	1.2
V_{max} (mM min^{-1})	4.6×10^{-4}
k_{cat} (min^{-1})	0.066
$k_{2\ (uncat)}$ (mM^{-1} min^{-1})	8.1×10^{-6}
$2k_{cat}/k_{2\ (uncat)}$ (M)	16

The kinetic parameters obtained for the bimolecular reaction are shown in Table 2. The K_m's for lactone and 1,4-phenylenediamine are 4.9 and 1.2 mM respectively; k_{cat} for the conversion of the two bound substrates to product is 0.066 min^{-1}. Obtaining a value for the rate acceleration due to the antibody is complicated in this case by the different units of the k_{cat} and the background rate constant k_2. The latter is second order, and has units of mM^{-1} min^{-1}. Thus, dividing k_{cat} by $k_{2(uncat)}$ gives a value with units of molar. In this case the value obtained is 16 M, which represents considerable acceleration over the background. Such a value is often known as the effective molarity, and it may be visualized as the concentration of one of the reactants required to bring the background rate up to the catalyzed rate. Note that the factor of 2 is to take account of the 2 equivalent amino groups in 1,4-phenylenediamine capable of background reaction.

In order to test whether catalysis of the bimolecular reaction was stereoselective, we separately assayed the reaction of each enantiomer of the lactone. Each was separately synthesized according to the scheme shown in Figure 6. A racemic sample of the hydroxy ester shown was derivatized with the (R)-enantiomer of α-methoxy mandelic acid. This gives two diastereomers ((R,R) and (R,S)) which, unlike the preceding pair of enantiomers, have different physical properties. These were separated by preparative thin layer chromatography. Hydrolysis of each ester gave a single enantiomer of the hydroxy-acid shown, which in turn gave a single enantiomer of the desired lactone. The ring-opening of each was assayed in the presence of antibody, and it was found that catalysis occurred in the case of the (R)-(-)-enantiomer only. Furthermore, the active isomer was shown by NMR to have the same configuration as that produced by the antibody-catalyzed cyclization reaction discussed earlier. This supports our expectation that the transition

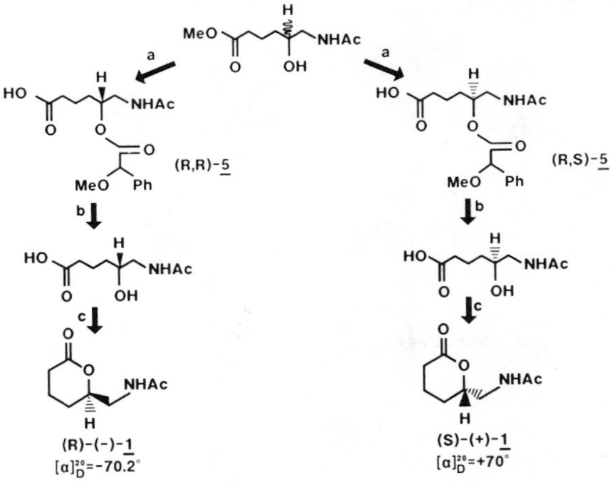

MeO NHAc O OH H
a a
HO NHAc O O O MeO Ph H (R,R)-5
HO NHAc O O O MeO Ph H (R,S)-5
b b
HO NHAc O OH H
HO NHAc O OH H
c c
O O NHAc H
(R)-(-)-1
$[\alpha]_D^{20}=-70.2°$
O O NHAc H
(S)-(+)-1
$[\alpha]_D^{20}=+70°$

Figure 6. Reagents and conditions: (a) (R)-(—)-α-methoxyphenylacetic acid, dicyclohexylcarbodiimide, N,N-dimethylaminopyridine, pyridine, 22 hr; (b) 1 M NaOH, 40 hr, 25°C; (c) aqueous HCl.

states of the cyclization and ring-opening are bound in the same orientation.

As already mentioned, the rationale behind our approach to generating catalytic antobodies was that binding should be tightest to the transition state of the reaction. We have already shown this qualitatively; the transition state analogue inhibitor binds more tightly than any of the substrates or products of either reaction. We may take this further, and predict a rate acceleration based on the difference in binding constants. A comparison of prediction and the actual observed rate accelerations is a good test of the original rationale.

The analysis is based on the principles of transition state theory. One of the central assumptions of this theory is that the transition state is in equilibrium with the ground state reactants. This allows construction of the scheme for a two substrate reaction, shown in Figure 7. In this case the substrates are designated L and A for lactone and amine respectively. The equilibrium constant for formation of the transition state from the free substrates is shown as $K_N‡$. Similarly, an equilibrium constant may be written for formation of the transition state from the antibody-bound substrates, shown as $K_{Ab}‡$. The scheme is completed by the equilibrium of binding of the substrates, given by the product of the binding constants K_L and K_A, and

$$\mathrm{Ab + L + A} \underset{}{\overset{K_N^‡}{\rightleftharpoons}} [\mathrm{LA}]^‡ + \mathrm{Ab} \longrightarrow \mathrm{P + Ab}$$

$$\Big\downarrow\!\!\uparrow K_L K_A \qquad\qquad \Big\downarrow\!\!\uparrow K_T$$

$$\mathrm{Ab.L.A.} \overset{K_{Ab}^‡}{\rightleftharpoons} [\mathrm{AbLA}]^‡ \longrightarrow \mathrm{P.Ab}$$

$$\text{Where:} \quad \frac{K_{Ab}^‡}{K_N^‡} = \frac{K_A K_L}{K_T} = \frac{k_{Ab}}{k_N}$$

Figure 7.

the binding constant of the transition state with antibody, $K_{T‡}$. From the scheme it is apparent that the ratio of $K_{Ab}^{‡}$ to K_N is equal to the ratio of K_LK_A to K_T. Transition state theory also allows the ratio of equilibrium constants $K_{ab}^{‡}/K_N^{‡}$ to be equated to the ratio of the corresponding reaction rate constants, shown here as k_{AB}/k_N. Thus, this analysis predicts that the rate acceleration due to the antibody is given by K_LK_A/K_T. K_L and K_A are the K_m values for the two substrates, and K_T may be approximated by the binding constant, K_i, of the transition state analogue. For a single substrate reaction the analysis is the same, except there is only the one substrate binding constant to consider. In other words, if the substrate is S, the rate acceleration is given by K_S/K_T.

TABLE 3.
COMPARISON OF PREDICTED TO ACTUAL
RATE ENHANCEMENTS

Bimolecular		Cyclization	
$\frac{k_{Ab}}{k_N}$	$\frac{K_{Ab}^{‡}}{K_N^{‡}}$	$\frac{k_{Ab}}{k_N}$	$\frac{K_{Ab}^{‡}}{K_N^{‡}}$
16 M	155 M	790	1000

Table 3 shows the extent to which this analysis agrees with our observations. In the case of the cyclization reaction the ratio of K_m of the hydroxy-ester substrate to K_i of the phosphonate inhibitor predicts a rate acceleration of approximately 1000. The observed value of 790 is in good agreement with this. For the bimolecular reaction the product of the substrate K_m's divided by the phosphonate K_i gives a value of 155 M. This compares with an experimentally determined effective molarity of 16 M. Thus, in this case the rate acceleration is significantly less than that predicted. However, this may reflect the additional para aromatic amino group in the bimolecular transition state, which is not present in the cyclization substrate or the transition state analogue. With both reactions it is clear that the rate acceleration may be accounted for by binding phenomena alone, without any need to invoke acid-base catalysis within the binding site of the antibody. In order to investigate this further we decided to study the pH dependence of the cyclization reaction. Evidence for or against acid-base catalysis is often obtained from pH rate profiles of enzyme catalyzed reactions. Thus we investigated the cyclization reaction in the pH range 6-9. Both the background and catalyzed reactions show a first order dependence on hydroxide ion concentration in this pH range (data not shown). This suggests that there is no acid-base catalysis due to a residue in the antibody-binding site. Furthermore, the same hydroxide ion catalyzed mechanism appears to operate in both cases.

The findings described here demonstrate that the precise predetermined binding specificity of immunoglobins may be used to generate catalysts which display the exquisite stereoselectivity characteristic of enzymes. The cyclization of 4 to 1, and the ring-opening of 1 with 1,4-phenylene-diamine are both catalyzed with near absolute enantioselectivity; also, as expected, the same enantiomer of 1 is processed in both reactions (9,10). This work, and the characterization of an antibody capable of stereospecific catalysis of a Claisen rearrangement (16), offers the promise of tailor-made, stereospecific catalysts for chemical processes.

ACKNOWLEDGMENTS

We are grateful to Diane Schloeder for expert technical assistance.

REFERENCES

1. Pressman D, Grossberg A (1968). "The Structural Basis of Antibody Specificity." New York: Benjamin.
2. Pauling L (1948). Chemical achievement and hope for the future. Am Sci 36:51.
3. Jencks W (1969). "Catalysis in Chemistry and Enzymology." New York: McGraw-Hill.

4. Wolfenden R, Frick L (1987). Transition state affinity and the design of enzyme inhibitors. In Page MI, Williams A (eds): "Enzyme Mechanisms," London: Royal Society of Chemistry, p 97.
5. Tramontano A, Janda KD, Lerner RA (1986). Chemical reactivity at an antibody binding site elicited by mechanistic design of a synthetic antigen. Proc Natl Acad Sci USA 83:6736.
6. Tramontano A, Janda KD, Lerner RA (1986). Catalytic antibodies. Science 234:1566.
7. Pollack SJ, Jacobs JW, Schultz PG (1986). Selective chemical catalysis by an antibody. Science 234:1570.
8. Jacobs J, Schultz PG, Sugasawara R, Powell M (1987). Catalytic antibodies. J Am Chem Soc 109:2174.
9. Napper AD, Benkovic SJ, Tramontano A, Lerner RA (1987). A stereospecific cyclization catalyzed by an antibody. Science 237:1041.
10. Benkovic SJ, Napper AD, Lerner RA (1988). Catalysis of a stereospecific bimolecular amide synthesis by an antibody. Proc Natl Acad Sci USA 85:5355.
11. Janda KD, Lerner RA, Tramontano A (1988). Antibody catalysis of bimolecular amide formation. J Am Chem Soc 110:4835.
12. Tramontano A., Ammann AA, Lerner RA (1988). Antibody catalysis approaching the activity of enzymes. J Am Chem Soc 110:2282.
13. Janda KD, Schloeder D, Benkovic SJ, Lerner RA (1988). Induction of an antibody that catalyzes the hydrolysis of an amide bond. Science 241:1188.
14. Hilvert D, Carpenter SH, Nared KD, Auditor MM (1988). Catalysis of concerted reactions by antibodies: the Claisen rearrangement. Proc Natl Acad Sci USA 85:4953.
15. Jackson DY, Jacobs JW, Sugasawara R, Reich SH, Bartlett PA, Schultz PG (1988). An antibody-catalyzed Claisen rearrangement. J Am Chem Soc 110:4841.
16. Hilvert D, Nared K (1988). Stereospecific Claisen rearrangement catalyzed by an antibody. J Am Chem Soc 110:5593.

Protein and Pharmaceutical Engineering, pages 105–118

CHEMICAL RESCUE AND CHANGE IN RATE-DETERMINING STEPS ELICITED BY SITE-DIRECTED MUTAGENESIS PROBES OF ASPARTATE AMINOTRANSFERASE[1]

Jack F. Kirsch, Michael D. Toney, and Jonathan M. Goldberg

Department of Biochemistry, University of California, Berkeley, California 94720

ABSTRACT Conversion of the aspartate aminotransferase active site residue, Lys258 to Ala (K258A) by site-directed mutagenesis yields an enzyme that is virtually inactive in transamination, but which binds α-amino and α-keto acids to produce aldimines and ketimines respectively. Enzyme activity is partially restored by the addition of primary amines to the reaction. These functionally replace the deleted ε-amino group of Lys258. A Brønsted β value of 0.4 is obtained for the chemical rescue effected by the series of eleven primary amines when steric effects are included in the regression analysis. The mutation Y225F removes the hydroxyl group which hydrogen bonds to the 3'-hydroxy function of the pyridoxyl cofactor. This mutant enzyme acting on aspartate has ketimine hydrolysis as its probable rate-determining step, while a combination of C_α-proton abstraction and oxalacetate dissociation is rate-determining for wild-type enzyme.

[1] This work was supported by NIH Grant GM 35393

INTRODUCTION

Aspartate aminotransferase (AATase) has become the paradigmatic enzyme for the study of vitamin B_6 catalyzed amino acid transformations. Its early crystallization (1) was followed by a series of successful experimental probes employing nearly every technique available to the mechanistic enzymologist. The important 1973 review by Braunstein (2), served to focus much of the work of the next decade which culminated in the first high-resolution crystal structure of aspartate aminotransferase (3,4). A detailed mechanistic proposal based on the combined information available from x-ray, kinetic and other structural data was proposed in 1984 (5).

The reaction catalyzed by this enzyme is shown in Equation 1.

$$H_2N\underset{|}{\overset{R_1}{C}}HCO_2H + (O)\overset{R_2}{\underset{|}{C}}CO_2H \rightleftharpoons H_2N\overset{R_2}{C}HCO_2H + (O)\overset{R_1}{C}CO_2H \quad (1)$$

The enzyme reacts preferentially with the dicarboxylic amino and keto acids (L-glutamate, L-aspartate, α-ketoglutarate, and oxalacetate), but will transaminate other amino acids, particularly those bearing aromatic side-chains, more slowly.

All of the chemistry takes place within the covalent confines of the pyridoxyl-substrate complex. Molecules related to pyridoxal phosphate (PLP) such as 3-hydroxy pyridine-4-aldehyde react with α-amino acids to yield the corresponding α-keto acids and the amine form of the coenzyme analog in the presence of divalent metals or of high concentration of buffer catalyst (6). Embedding the coenzyme within the protein superstructure increases the catalytic efficacy by approximately 10^7(7). It is reasonable to start with the hypothesis that the protein functional groups most closely associated with the pyridoxyl derivative are the most important in influencing the catalytic transformation. The problem facing the mechanistic enzymologist is how to factor this large catalytic rate acceleration, as well as the qualitative features of the mechanism, into the individual contributions of the amino acids surrounding the cofactor. Site-directed mutagenesis offers the surgical precision necessary to address this question. Some of the early successes as well as the limitations of this approach have been reviewed recently by Gerlt (8) and by Knowles (9).

The roles of Arg292 (10) and Tyr70 (11) in determining respectively the substrate specificity and the affinity of the enzyme for the coenzyme, have been addressed in earlier publications from this laboratory. An overall summary has also appeared (12). The present report addresses some of the properties of the mutants K258A and Y225F. An abbreviated mechanism is given in Scheme 1. The detailed proposal has been discussed elsewhere (4,5). The scheme illustrates the spatial relationship of K258 and Y225 to the cofactor, and emphasizes the cardinal role of the Lys ε-NH_2 group in catalyzing the crucial step in enzymatic transamination — namely the 1,3 azaallylic rearrangement separating aldimine from ketimine.

CHEMICAL RESCUE — THE LAZARUS EFFECT

The mutant K258A reacts in its pyridoxamine phosphate form (PMP) with α-ketoglutarate to give the ketimine (12). Further progress to the aldimine has a half life of many hours. The PLP form of the K258A enzyme correspondingly reacts with amino acids to yield an aldimine which is converted to the ketimine very slowly because of the deletion of the critical ε-NH_2 of K258A. Transamination in either direction is thus barely detectable under normal conditions.

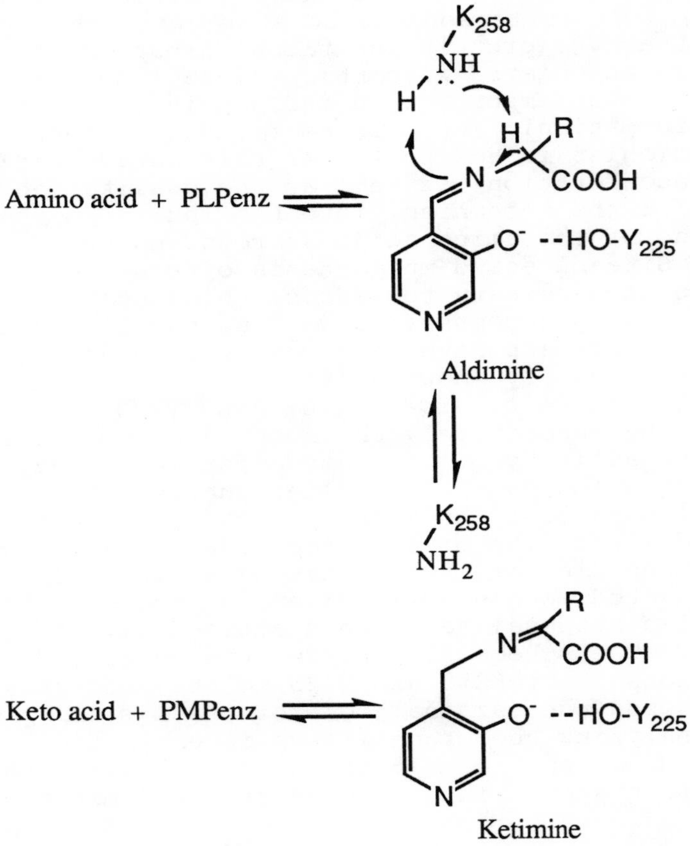

Amino acid + PLPenz ⇌ Aldimine

Aldimine ⇌ Ketimine

Keto acid + PMPenz ⇌ Ketimine

SCHEME 1.

As conversion of Lys258 to Ala removes $-(CH_2)_3-NH_2$ from the active site of the enzyme, the question arose as to whether the catalytically essential amino group might be functionally replaced by the addition of simple aliphatic amines to the external aldimine form of K258A. The recently solved x-ray structure of this mutant shows only minor and local changes existing between K258A and wild-type enzyme, with the major change specifically at the site of the mutation (S. Almo, D. Smith, M. Toney, and D.Ringe, unpublished results) The concept is diagramed in Scheme 2.

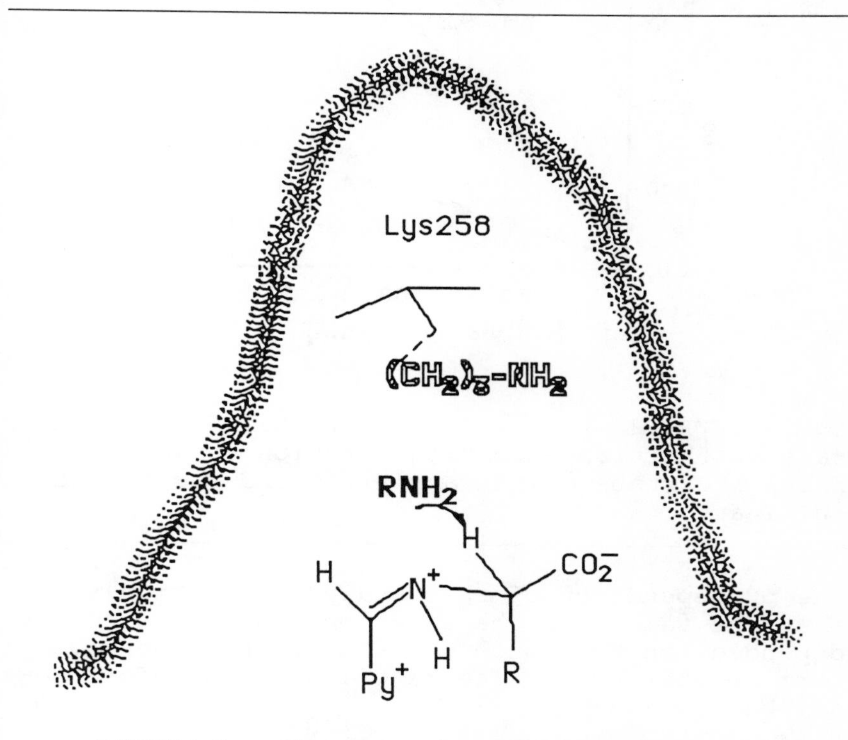

SCHEME 2. The mutation K258A deletes that part of the side chain of Lys258 shown in outline text. The released space could be occupied by added amines (bold face) which catalyze the transamination.

Cysteine sulfinic acid, a reactive analog of aspartate, combines readily with the PLP form of K258A to produce the external aldimine (7). Transamination is significantly accelerated by a variety of primary amines. A typical plot of the increase in the rate constant for transamination as a function of the concentration of added amine is shown in Figure 1.

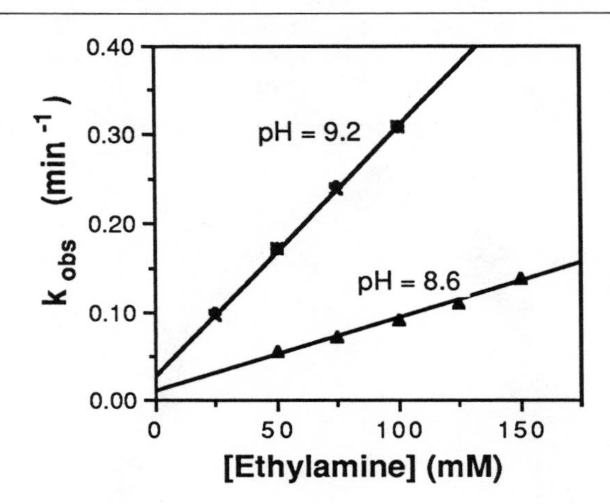

FIGURE 1. The effect of ethylamine on the rate of the transamination reaction between *E. coli* K258A aspartate aminotransferase and L-cysteine sulfinate.

The pH dependence of the catalyzed reaction shows that the values of the rate constant are linearly dependent on the concentration of the free-base form of the amine. The rate law is therefore described by Equation 2.

$$k_{obs} = \frac{k_B \ [\text{amine}]_{total}}{1 + ([H^+]/K_a)} + k_{solvent} \tag{2}$$

The nearly dead enzyme, K258A, can thus be rescued by the addition of amine catalysts which replace the deleted functionality of the enzyme. This result offered the possibility of conducting a true Brønsted analysis of proton transfer for an enzyme catalyzed reaction. A Brønsted plot is a measure of the sensitivity of the rate constant for a reaction to the pK_a of the catalyzing acid or base (13) (Equation 3).

$$\log k_B = \beta pK_a + \text{const.} \qquad (3)$$

The slope of the line so generated is often interpreted as a measure of the degree of charge development, or of proton transfer, in the transition state for the reaction. Brønsted analysis is thus an important measure of transition state structure, but its application to enzymology has heretofore been precluded by the fact that the catalyzing acids or bases are integral parts of the enzyme structure.

A series of eleven primary amines was investigated as catalysts for the transamination of cysteine sulfinate (7). Negative steric interaction factors are significant within this series as might have been expected from consideration of the crowded architecture of an active site. There is a decrease of 15-fold/CH_2 in the amine catalyzed rate constant over the series methyl, ethyl, propyl, and butyl amine. The steric effects are cleanly isolated in this limited series of four amines, as all members have nearly the same pK_a. The rate constants generated by the entire family of 11 primary amines whose pK_a's range from 5.3-10.6 are well accommodated by the multiple linear model shown in Equation 4.

$$\log k_B = \beta(pK_a) + V\,(\text{molec. vol.}) + c \qquad (4)$$

where both pK_a and solvent-excluded molecular volume are the independent, and β, V, and c are the adjustable parameters. A plot analogous to the conventional Brønsted presentation, but corrected for the steric effects, is generated by a graph of $\log k_B$-(V x molec. vol.) *vs.* pK_a and is shown in Figure 2. The least squares minimizing values of β and V are 0.39 ± 0.05 and (-0.055 ± 0.005) $Å^3$, respectively. The value of β is interpreted in

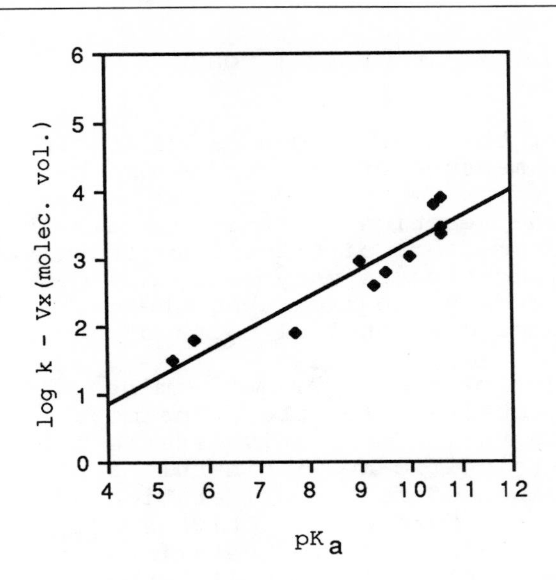

FIGURE 2. Brønsted-type plot for the acceleration of transamination between aspartate aminotransferase and cysteine sulfinate by added amines. The slope of the line, the β value, is 0.4. The amines and their pK_a values are: methyl- (10.6), ethyl- (10.6), propyl- (10.5), butyl- (10.6), ethylenedi- (10.0), ethanol- (9.5), ammonia (9.2), 2-fluoroethyl- (9.0), 2-cyanoethyl- (7.7), 2,2,2-trifluoroethyl- (5.7), cyanomethyl- (5.3). (From (7).)

terms of the development of approximately 40% positive charge on the accepting amine with a corresponding 40% carbanionic character on the PLP adduct in the transition state. It should be possible to apply a similar approach to other enzymes where the location of catalytic acidic or basic groups permits access of replacement molecules from solvent.

CHANGES IN RATE-DETERMINING STEPS

Site-directed mutagenesis of Tyr225 to Phe (Scheme 1) yields an AATase with several significantly altered properties (Goldberg and Kirsch unpublished results). A prominent feature is that the values of k_{cat}/K_m and k_{cat} measured in the aspartate to oxalacetate direction are reduced 20- and 450-fold, respectively, in this mutant. The question to be addressed here is: Is the decrement in rate constants due primarily to a decrease in the microscopic rate constants which are overall rate-determining for the wild-type enzyme or do some steps which are less significant kinetically for wild-type enzyme become rate-determining for the mutant? These considerations are diagramed in Figure 3.

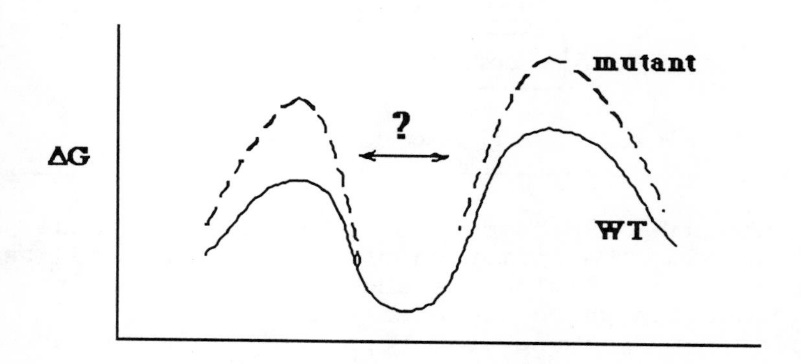

FIGURE 3. Is the decrease in a complex rate parameter (*i.e.* V or V/K) due to a mutation, the result of a decrease in a rate constant that originally limits the velocity of the wild-type enzyme (right solid peak), or of a greater decrease in a rate constant that is not kinetically significant for wild-type enzyme (left peaks)? This hypothetical wild-type profile (solid line) is rate-limited by the second step. A mutation might further slow the original rate-determining step (right dashed peak),or cause a new step to become limiting (left dashed peak).

The AATase half reaction must be minimally comprised of the five steps shown in Scheme 3. The first and last steps are simple diffusive processes

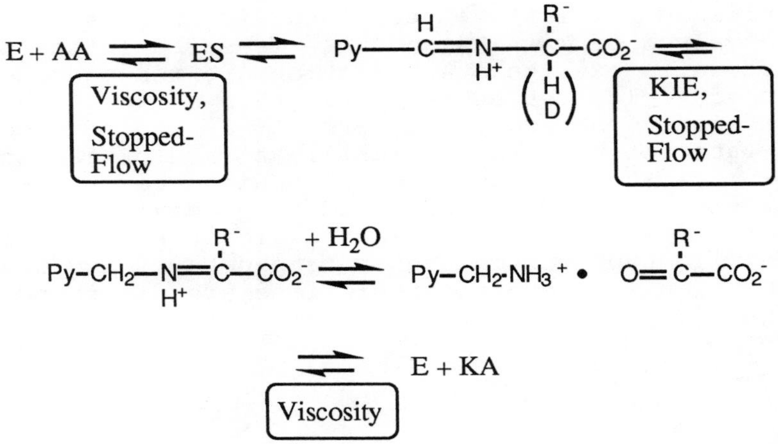

E + AA ⇌ ES ⇌ Py—CH=N(H^+)—C(R$^-$)(H/D)—CO_2^- ⇌

[Viscosity, Stopped-Flow] [KIE, Stopped-Flow]

Py-CH_2—N(H^+)=C(R$^-$)—CO_2^- + H_2O ⇌ Py–CH_2-NH_3^+ • O=C(R$^-$)—CO_2^-

⇌ E + KA

[Viscosity]

SCHEME 3.

which would be expected to be sensitive to viscosity (14,15). The formation of the aldimine and its conversion to ketimine are accompanied by spectroscopic changes which are easily accessible by stopped-flow spectrophotometry. The aldimine to ketimine conversion involves additionally the labilization of the C_α-H bond, and should therefore exhibit an α-hydrogen kinetic isotope effect. Unfortunately, there is no easy way to monitor the hydrolysis of the ketimine to enzyme bound α-keto acid and pyridoxamine phosphate.

Aldimine to ketimine conversion is at least partly rate-determining for wild-type enzyme as evidenced by the ^DV = 1.9 (Figure 4). This kinetic isotope effect vanishes with the mutant Y225F, showing definitively that the *relative* peak heights in the free energy diagram describing the reaction coordinate profile of the mutant and wild-type enzymes differ, and that the effect of the mutation is as described on the left side of Figure 3.

Both V and V/K for aspartate are decreased by increasing viscosity due to added sucrose or glycerol (Figure 5). (The glycerol data are not shown.) Such a result is interpreted in terms of a partly rate-determining association of substrate (14,15) or dissociation of product (16). The

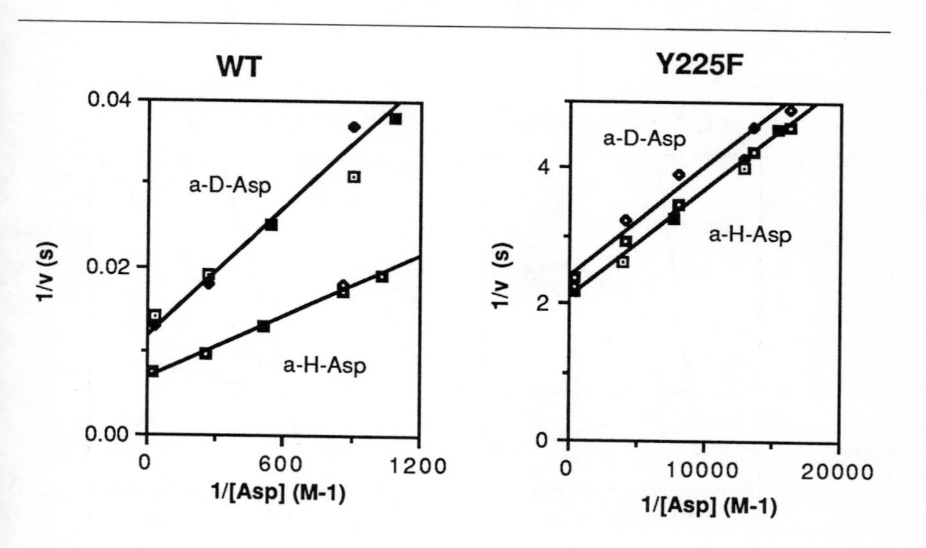

FIGURE 4. The primary C_α-hydrogen kinetic isotope effect on the AATase catalyzed transamination of aspartate. Conditions: pH 9.0; 25°C.

effect on V as well as on V/K implies that the viscosity experiments are a measure of the diffusion of oxalacetate from the enzyme bound PMP-oxalacetate complex, rather than of the association of aspartate with the enzyme and that the former is partly rate-determining for wild-type enzyme. (The complex constant, V, measures steps occuring only after ES complex formation.) Caution must be maintained in the interpretation of such experiments, because of the possibility of non-specific effects on the enzyme caused by the high concentrations of the viscosogenic reagent. This effect is ordinarily controlled by the use of a more slowly reacting control substrate (14,15,17). But here, as in the case of triosephosphate isomerase (18), the more slowly reacting mutant enzyme serves as an appropriate control. The effect of viscosity on the kinetic parameters of Y225F is much less than it is on those for wild-type enzyme implying that dissociation of oxalacetate is not rate-determining for the mutant. The species with an absorbtion maximum of 430nm, probably corresponding to the formation of the aldimine from the Y225F-Asp

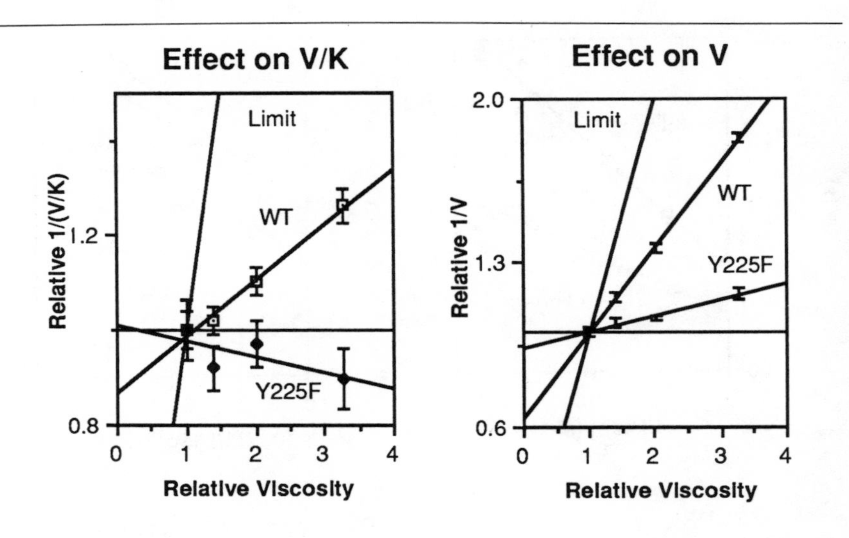

FIGURE 5. The effects of increasing viscosity on the values of V and V/K for the AATase catalyzed transamination of aspartate. Conditions: pH 9.0; 25°C.

complex rises with a limiting rate constant of ca. 150 s^{-1}, which is very much greater than the k_{cat} value of 0.4 s^{-1} (Goldberg and Kirsch, unpublished observations). The remaining step in the minimal mechanism shown in Scheme 3, ketimine hydrolysis, is thus, by elimination, the most likely candidate for the assignment of the rate-determining step for the mutant, Y225F. The rate-determining step for wild-type enzyme, on the other hand, probably represents a combination of the rate constants for C_α-H abstraction, oxalacetate dissociation, and possibly ketimine hydrolysis.

REFERENCES

1. Jenkins WT, Sizer IW (1957). Glutamic aspartic transaminase. J Am Chem Soc 79:2655.
2. Braunstein AE (1973). Amino group transfer. Enzymes 9:379.
3. Ford GC, Eichele G, Jansonius JN (1980). Three-dimensional structure of a pyridoxal-phosphate-dependent enzyme, mitochondrial aspartate aminotransferase. Proc Natl Acad Sci USA 77:2559.
4. Jansonius JN, Vincent MG (1987). Structural basis for catalysis by aspartate aminotransferase. In Jurnak FA, McPherson A (eds):"Biological Macromolecules and Assemblies," vol 3, New York: Wiley, p. 187.
5. Kirsch JF, Eichele G, Ford GC, Vincent MG, Jansonius JN, Gehring H, Christen P (1984). Mechanism of action of aspartate aminotransferase proposed on the basis of its spatial structure. J Mol Biol 174:497.
6. Auld DS, Bruice TC (1967). Catalytic reactions involving azomethines. IX. General base catalysis of the transamination of 3-hydroxypyridine-4-aldehyde by alanine. J Am Chem Soc 89:2098.
7. Toney MD, Kirsch JF (1989). Direct Brønsted analysis of the restoration of activity to a mutant enzyme by exogenous amines. Science in press.
8. Gerlt JA (1987). Relationships between enzymatic catalysis and active site structure revealed by applications of site-directed mutagenesis. Chem Rev 87:1079.
9. Knowles JR (1987). Tinkering with enzymes: What are we learning? Science 236:1252.

10. Cronin CN, Kirsch JF (1988). Role of arginine-292 in the substrate specificity of aspartate aminotransferase as examined by site-directed mutagenesis. Biochemistry 27:4572.
11. Toney MD, Kirsch JF (1987). Tyrosine 70 increases the coenzyme affinity of aspartate aminotransferase. A site-directed mutagenesis study. J Biol Chem 262:12403.
12. Kirsch JF, Finlayson WL, Toney MD, Cronin CN (1987). Mechanistic analysis of the aspartate aminotransferase active site mutants -- Y70F, K258A, and R292D. In Korpela T, Christen P (eds): "Biochemistry of Vitamin B6, Proceedings of the 7th International Congress on Chemical and Biological Aspects of Vitamin B6 Catalysis," Basel: Birkhäuser, p 59.
13. Jencks, WP (1969). "Catalysis in Chemistry and Enzymology." New York: McGraw-Hill, chap.3.
14. Brouwer AC, Kirsch JF (1982). Investigation of diffusion-limited rates of chymotrypsin reactions by viscosity variation. Biochemistry 21:1302.
15. Hardy LW, Kirsch JF (1984). Diffusion-limited component of reactions catalyzed by Bacillus cereus ß-lactamase I. Biochemistry 23:1275.
16. Stone SR, Morrison JF (1988). Dihydrofolate reductase from *Escherichia coli*: the kinetic mechanism with NADPH and reduced acetylpyridine adenine dinucleotide as substrates. Biochemistry 27:5493
17. Bazelyansky M, Robey E, Kirsch JF (1986). Fractional diffusion-limited component of reactions catalyzed by acetylcholinesterase. Biochemistry 25:125.
18. Blacklow SC, Raines RT, Lim WA, Zamore PD, Knowles JR (1988). Triosephosphate isomerase catalysis is diffusion controlled. Biochemistry 27:1158.

Protein and Pharmaceutical
Engineering, pages 119–134

Molecular Approaches in Analysis of the Substrate Specificity of Trypanothione Reductase, a Flavoprotein from Trypanosomatid Parasites[1]

Francis X. Sullivan†, R. Luise Krauth-Siegel§, Emil F. Pai‡, and Christopher T. Walsh†

†Department of Biological Chemistry and Molecular Pharmacology, Harvard Medical School, Boston, MA 02115

§Institut fur Biochemie II, Medizinische Fakultat der Unversitat, Im Neuenheimer Feld 328, D-6900, Heidelberg, FRG

‡Abteilung Biophysik, Max Planck-Institut fur Medizinische Forschung, Jahnstr.29, D-6900, Heidelberg, FRG

ABSTRACT. Trypanothione reductase catalyzes the NADPH dependent reduction of trypanothione, N^1,N^8 bis(glutathionyl) spermidine, a glutathione analog unique to the trypanosomatid parasites. The enzyme is structurally and mechanistically very similar to the host enzyme human glutathione reductase. However, the host and parasite enzyme show mutually exclusive substrate specificity. This has made trypanothione reductase a promising target for the rational design of trypanocidal agents. Unfortunately, little structural or mechanistic information has been available on the reductase. The trypanothione reductase gene from *T. congolense* has now

[1] This work was in part supported by a grant from the National Institute of Health, GM 21643, and a postdoctoral fellowship from the Damon Runyon-Walter Winchel Cancer Fund (FXS).

been cloned and expressed in *E. coli*. Here we report crystallization of the *T. congolense* enzyme and describe several initial site directed mutants that probe the reductase's structure and examine residues that may be involved in substrate binding. In particular, the C-terminal twenty amino acids of trypanothione reductase appear to be necessary for catalysis of the disulfide reduction half reaction and two residues, glutamic acid 18 and tryptophan 21, appear to be involved in binding trypanothione. These later two residues are analogous to ones implicated in substrate binding in human glutathione reductase from the high resolution crystal structure of that enzyme.

DISCOVERY OF TRYPANOTHIONE

During their studies on glutathione metabolism of the insect trypanosomatid *Crithidia fasciculata*, Cerami and coworkers isolated a heat stable compound required for glutathione reductase activity (1). They identified this compound as the glutathione analog N^1,N^8, bis(glutathionyl)spermidine, trivially called trypanothione (See figure 1). Trypanothione was later shown to be present in all trypanosomatid parasites examined (2). The subsequent discovery that trypanosomes had replaced glutathione dependent peroxidase with a trypanothione dependent peroxidase (3) strengthened the view that the parasites had evolved a metabolism based in part upon this novel glutathione analog and that this could be exploited in the rational design of trypanocidal agents. Given the role of glutathione and the glutathione reductase - glutathione dependent peroxidase redox couple in the cell's response to oxidative stress (eqn 1 and 2)(4) and the trypanosomatid parasites' demonstrated oxygen sensitivity (5) this seemed to be a particularly advantageous strategy.

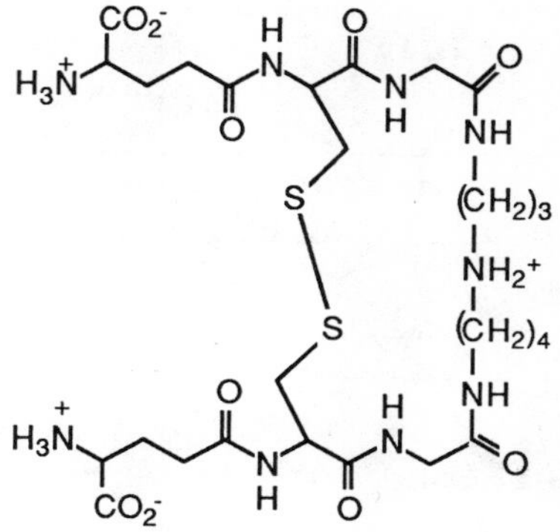

Figure 1. Oxidized trypanothione.

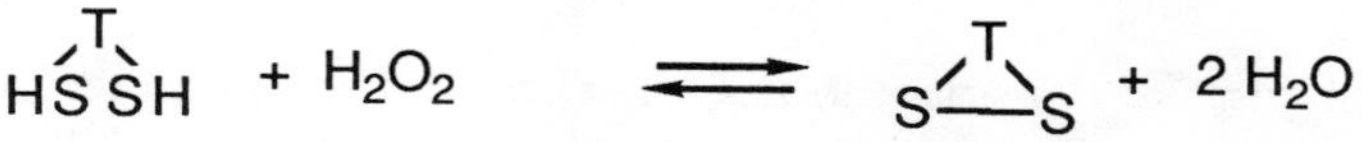

$$T(SH)_2 + H_2O_2 \rightleftharpoons TS_2 + 2\,H_2O \qquad (1)$$

Trypanothione Peroxidase

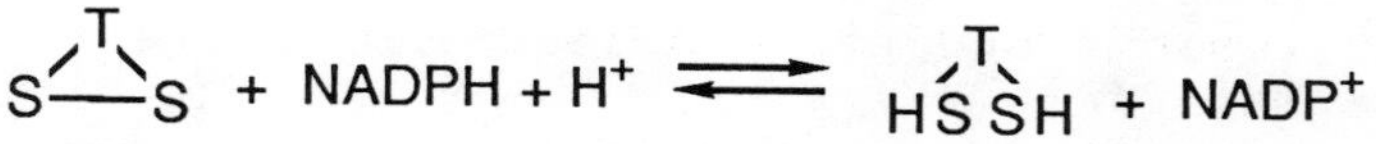

$$TS_2 + NADPH + H^+ \rightleftharpoons T(SH)_2 + NADP^+ \qquad (2)$$

Trypanothione Reductase

TRYPANOTHIONE REDUCTASE

Purification of Trypanothione Reductase

Trypanothione reductase was first purified from *C. fasciculata* by Shames et al.(6) and shown to be a member of the glutathione reductase family of disulfide containing flavoproteins. Subsequently Krauth-Siegel et al.(7) purified the reductase from the human parasite *T. cruzi*. Trypanothione reductase catalyses the NADPH dependent reduction of trypanothione (eqn 2). The reductases from both *C. fasciculata* and *T. cruzi* are very similar to glutathione reductase, both physically and mechanistically. They have similar monomer molecular weights, 50-55 kD, and a similar active dimeric structure. Each contains an FAD cofactor and a reducible active site

TABLE 1.
SUBSTRATE SPECIFICITY OF TRYPANOTHIONE REDUCTASE VS. GLUTATHIONE REDUCTASE

Turnover number (min-1)	Trypanothione Reductase[a]	Glutathione Reductase[b]
Glutathione	3.1[c]	12,000.
Trypanothione	31,000.	9.6[d]

a) *C. fasciculata* b) Human erythrocyte
c) 50 mM Glutathione d) 0.28 mM Trypanothione

disulfide as key redox components. Each utilizes NADPH as a source of reducing equivalents. They have similar optical spectra and form charge transfer complexes between one active site cysteine residue and the oxidized FAD in the enzyme's EH_2 form.

However, the parasite and host reductases differ dramatically in their substrate specificity (Table 1). Trypanothione reductase will not reduce glutathione and glutathione reductase will not reduce trypanothione, at any significant rate. Since the host and parasite enzymes show mutually exclusive substrate specificity, trypanothione reductase appears as an intriguing target for the rational design of trypanocidal agents based upon the selective inhibition of its activity. Unfortunately, little structural information was available on the protein to guide inhibitor design. Krauth-Siegel et al.(7) had grown microcrystals of trypanothione reductase from *T. cruzi* but were limited by the amount of available enzyme. Thus, to obtain some primary sequence information and to circumvent the problems associated with isolating a protein present only in small amounts in a pathogenic organism, Shames et al.(8) recently cloned the gene encoding trypanothione reductase.

Cloning and Sequencing of the Trypanothione Reductase Gene from *T. congolense*

The trypanothione reductase gene from *T. congolense*, an African cattle pathogen, was cloned and sequenced by Shames et al.(8). The primary sequence confirmed the functional similarities between trypanothione reductase and glutathione reductase. Furthermore, it indicated that trypanothione reductase was similar in structure to the other members of the family of FAD containing oxidoreductases including glutathione reductase, mercuric reductase (9), and lipoamide dehyrogenase (10). Overlaying the primary sequence of trypanothione reductase onto the high resolution crystal structure of human glutathione reductase (11,12,13,14) revealed that many of the essential structural features and most of the catalytically important residues were most likely conserved between the host and parasite enzyme (Table 2). Structurally conserved were the FAD binding site, the NADPH binding site, the subunit interface domains and several

TABLE 2.
COMPARISON OF KEY RESIDUES IN HUMAN ERYTHROCYTE GLUTATHIONE REDUCTASE TO *T. CONGOLENSE* TRYPANOTHIONE REDUCTASE[a]

Glutathione Reductase	Function from X-ray Structure	Trypanothione Reductase
$GGGSGGL_{35}$	FAD binding site	$GAGSGGL_{47}$
Cys - 58	Redox active disulfide	Cys - 52
Cys - 63	Redox active disulfide	Cys - 57
Lys - 57	Binds carboxylate of GS_1 γ-Glu moiety	Lys - 61
Tyr - 114	stacks between GS moieties in glutathione	Tyr - 110
His - 467	Active site base	His - 461
Glu - 472	H bonds to active site His	Glu - 466
Arg - 37	Binds carboxylate of GS_1 glycyl carboxylates via H bonds	Trp - 21
---------	Unknown	20 aa tail

a) From Shames et al., 1988.

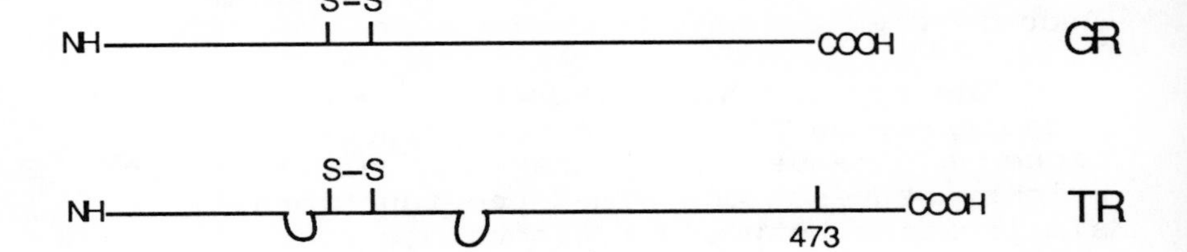

FIGURE 2. Comparison of *E. coli* glutathione reductase (GR) to *T. congolense* trypanothione reductase (TR).

residues assigned catalytic roles from the crystal structure of human glutathione reductase.

On closer inspection, several differences between trypanothione reductase and glutathione reductase that could potentially affect the substrate specificity were noted. These included the substitution for arginine 37 in glutathione reductase by a tryptophan at position 21 in trypanothione reductase. In glutathione reductase this arginine contacts one of the glycyl carboxylates of glutathione. In addition, trypanothione reductase possesses a twenty amino acid C-terminal extension relative to the sequence of glutathione reductase (Figure 2). A similar C-terminal extension present in mercuric reductase has recently been shown to help comprise the active site of that enzyme and to be essential for catalysis (15,16). *T. congolense* trypanothione reductase also contains two peptide insertions, of nine and twelve residues, relative to the sequence of glutathione reductase. The nonapeptide insertion lies just prior to the active site cysteines. This peptide insertion is present in both the *T. congolense* and *C. fasciculata* enzymes. Both insertions lie at positions analogous to a turn in the structure of glutathione reductase. This is represented diagrammatically in Figure 2.

Expression of *T. congolense* Trypanothione Reductase

We have recently expressed the trypanothione reductase gene from *T. congolense* in *E. coli* at a level of 1% of the soluble protein (Sullivan et al., submitted). This has allowed the facile purification of tens of milligrams of reductase, providing a safe, convenient source of the enzyme and solving the problem of enzyme availability. The reductase expressed in *E. coli* is by all criteria a representative member of the trypanothione reductase family. The subunit molecular weight is 54 kD, as predicted from the gene sequence. The enzyme has a dimeric structure, contains FAD, and utilizes NADPH. The *T. congolense* enzyme has optical spectra similar to the other trypanothione reductases and forms the characteristic charge transfer complex between the reduced flavin and the active site cysteine disulfide. Perhaps most importantly, it shows the same strict substrate specificity as the other trypanothione reductases. The availability of large amounts of *T. congolense*

Figure 3. Crystals of *T. congolense* trypanothione reductase.

trypanothione reductase expressed in *E. coli* and the availability of the cloned gene and its sequence for attempts at site directed mutagenesis should further investigation into the structure of this enzyme.

Obviously a crystal structure would benefit the rational design of potent inhibitors of trypanothione reductase, e.g. mechanism based inactivators and slow binding inhibitors. Recently we have obtained initial crystals of *T. congolense* trypanothione reductase (Figure 3).

Site Directed Mutagenesis of Trypanothione Reductase

Currently we are using the high resolution crystal structure of glutathione reductase with substrate bound (13) to model the interactions on trypanothione reductase that may be responsible for substrate discrimination. Initially we have focussed our attention on two regions of the reductase, the C-terminal twenty amino acid tail of trypanothione reductase and the residues on trypanothione reductase analogous to those on glutathione reductase that contact the glycyl carboxylates of glutathione. As seen in figure 4 the glycyl carboxylates of glutathione form amide bonds in trypanothione, converting a bis anion species to a monocation macrocyclic disulfide.

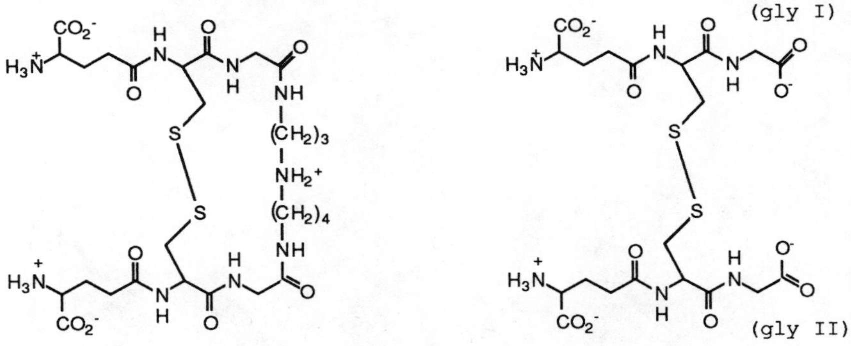

CO_2^- H O O
$H_3\overset{+}{N}$ N N O
O H NH
S $(CH_2)_3$
NH_2^+
S $(CH_2)_4$
+ O H NH
H_3N N N O
CO_2^- H O

Oxidized Trypanothione

(gly I)
CO_2^- H O O
$H_3\overset{+}{N}$ N N
O H O^-
S
S
+ O H O^-
H_3N N N O
CO_2^- H O
(gly II)

Oxidized Glutathione

Figure 4. Trypanothione and glutathione.

The work of Henderson et al. (17) has shown that these carboxylates are required for efficient binding of glutathione to glutathione reductase.

The residues on glutathione reductase that interact with the glycyl carboxylates of glutathione disulfide are shown in Table 3. There are two residues in glutathione reductase that contact the glycyl carboxylates in glutathione I (one of the glutathione units in glutathione disulfide) that are substituted in *T. congolense* trypanothione reductase. They are alanine 34, which is glutamic acid 18 in trypanothione reductase, and arginine 37, which is tryptophan 21 in trypanothione reductase. There are also two residues on glutathione reductase which interact with the glycyl carboxylate of glutathione II that are substituted in trypanothione reductase. However since glutathione II appears poorly bound in the crystal structure of glutathione reductase and since similar changes are present in *E. coli* glutathione reductase we have concentrated our initial efforts on the residues interacting with glutathione I. Since interactions with the glycyl carboxylate of glutathione might conceivably play a major role in substrate

TABLE 3.
RESIDUES CONTACTING GLYCYL CARBOXYLATES OF BOUND GLUTATHIONE IN THREE HOMOLOGOUS REDUCTASES

Substrate Group Glutathione	Protein Group Human GR	Residue Change Eco GR	TR
Gly I	Ala 34 (main)	---	Glu
	Arg 37 (side)	Asn	Trp
	Tyr 114(side)	---	---
Gly II	Ile 113(side)	Ser	Ser
	Tyr 114(side)	---	---
	Asn 117(side)	Val	Met

GR - Glutathione Reductase
TR - Trypanothione Reductase
Adapted from Karplus et al., (13).

discrimination, it seemed possible that substitutions of the residues on trypanothione reductase making these contacts could change the substrate specificity of the enzyme.

Using site directed mutagenesis we have replaced the two residues on trypanothione reductase analogous to alanine 34 and arginine 37 of glutathione reductase with an alanine (E18A) and an arginine (W21R). We also truncated trypanothione reductase to remove the C-terminal twenty amino acid extension by changing threonine 473 to a stop codon (T473*) (See Figure 2). Finally we constructed a triple mutant having all three changes (E18A W21R T473*). The mutant proteins were expressed in *E. coli* strain SG5 (18), a glutathione reductase deletion mutant. Selective labeling of the plasmid encoded reductases with ^{35}S methionine (19) revealed that each mutant reductase was expressed at levels similar to the wild-type enzyme and had similar stabilities *in vivo* (Data not shown). Crude extracts of each mutant were assayed for transhydrogenase activity with NADPH and thioNADP. This tests the enzyme for competency in reduction of the flavin in the first half reaction of disulfide reduction and demonstrates

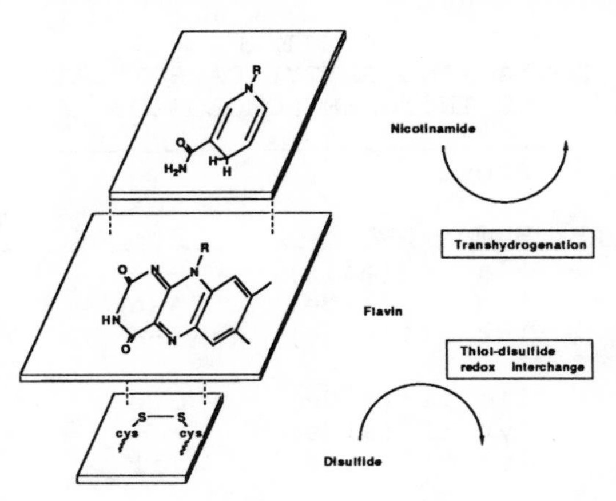

FIGURE 5. Schematic of active site of trypanothione reductase. Adapted from Pai and Schulz (12).

that the reductase is probably folded properly. Each mutant was also assayed for disulfide reductase activity with glutathione and the substrate analog deazatrypanothione. Due to limited amounts of deazatrypanothione we could measure the activity of each reductase at only one substrate concentration. Therefore, at this time we can not say whether the mutations affect the enzyme's K_M, k_{cat} or both.

The mutation that deleted the twenty amino acid tail removed the disulfide reductase activity (Table 4). Since this enzyme was stable in *E. coli*, and was competent in transhydrogenation, it appears that the protein was most likely folded correctly, and at least one half of the active site was intact. It can be inferred from these data that the C-terminal 20 amino acids of trypanothione reductase are involved in forming the active site and in substrate binding. This is analogous to the situation with mercuric reductase where two cysteines in the C-terminal extension are involved in catalysis (15,16).

TABLE 4.
ASSAYS FOR CATALYTIC ACTIVITY OF TRUNCATED T.CONGOLENSE TR (T473*)(IN CRUDE EXTRACTS)

Assay	WT TR (493aa)	Truncated TR (473aa)	WT GR (480aa)
A) Transhydrogenation: NADPH + thioNADP $\rightleftharpoons$ thioNADPH + NADP	100%	75%	100%
B) Trypanothione Reduction: $T(S)_2$ + NADPH $\rightleftharpoons$ NADP + $T(SH)_2$	100%	0%	0%
C) Glutathione Reduction: GSSG + NADPH $\rightleftharpoons$ NADP + 2GSH	0%	0%	100%

TABLE 5.
ACTIVITIES OF MUTANT TR

Mutant	Trans-hydrogenase Activity % of WT	Trypanothione Reductase Activity % of WT	Glutathione Reductase Activity
W 21 R	103	11	none
E 18 A	100	53	none
T 473 *	75	0	none
E 18 A W 21 R T 473 *	78	0	none

The changes at Trp 21 and Glu 18, the residues on trypanothione reductase, analogous to those residues that contact the glycyl carboxylates of glutathione$_I$ in glutathione reductase, led to reduced activity with trypanothione but did not increase activity with glutathione. Thus it appears these contacts are significant for trypanothione binding, but exchanging them was not sufficient to change the substrate specificity of the reductase.

The above experiments represent the initial attempts to delineate the residues involved in substrate discrimination. Obviously, the experiments presented are preliminary in nature, and need to be repeated with purified enzyme. In addition, in order to make more definitive statements, we need to examine more mutants and to examine each mutant more thoroughly with a variety of substrate analogs. But even at the present level of analysis we have been able to gain some insight into the structure of trypanothione reductase and the residues responsible for substrate binding.

CONCLUSION

The discovery of trypanothione and subsequently the trypanothione reductase - trypanothione dependent peroxidase redox couple has provided a new avenue for the rational design of trypanocidal agents. The cloning of the trypanothione reductase gene from *T. congolense* has provided the first structural information on the reductase and its expression in *E. coli* has overcome problems associated with isolating the enzyme from its natural source, providing ample quantities of the enzyme for further research. An X-ray structure of trypanothione reductase is actively being pursued and attempts at crystallization have met with some initial success. Site directed mutagenesis has allowed for initial exploration of residues involved in substrate binding and has yielded preliminary data on a structural feature of the enzyme. Further study along these lines should lead to a more complete picture of the mechanism of substrate discrimination employed by trypanothione reductase and greatly aid the design of specific inhibitors of the parasite enzyme.

EXPERIMENTAL

Materials and Strains.

Deazatrypanothione was a generous gift of Dr. G. Henderson, (The Rockefeller University, New York, NY). NADPH and thioNADP were obtained from Boehringer-Mannheim. All other biochemical reagents were obtained from Sigma Chemical Co. *E.coli* strain SG5 (18), a glutathione reductase deletion mutant, was the gift of Dr. R. Perham (Cambridge University, Cambridge, England).

Site Directed Mutagenesis.

Site directed mutagenesis was performed upon *T. congolense* trypanothione reductase expression vector pT7TR-1 (Sullivan et al, submitted) with the Amersham site directed mutagenesis kit according to the manufacturer's instructions. Performing mutagenesis directly upon the phagemid vector pT7TR-1 obviated the need to reconstruct the mutants.

Expression of *T. congolense* Trypanothione Reductase in *E. coli*.

T. congolense trypanothione reductase was expressed in *E. coli* glutathione reductase mutant SG5, to eliminate any background contamination from host glutathione reductase, according to Sullivan et al, (submitted). Briefly this involved expression of the reductase from a T7 RNA polymerase promoter containing plasmid,pT7TR-1, as part of a two plasmid system (19) induced by incubation for 2 hrs at 42 °C.

Purification and Crystallization of *T. congolense* Trypanothione Reductase.

T. congolense trypanothione reductase was purified to homogeneity as described in Sullivan et al (submitted). The enzyme was crystallized using a procedure slightly modified from that used for the *T. cruzi* enzyme (7).

Enzyme Assays.

Assays were performed as described in Shames et al.(6). The disulfide reductase assays contained either 80 μM deazatrypanothione or 20 mM glutathione.

ACKNOWLEDGMENTS

We would like to thank Drs G. Henderson and R. Perham for their generous gift of materials.

REFERENCES

1. Fairlamb AH, Blackburn P, Ulrich P, Chait BT, Cerami, A (1985). Trypanothione: a novel bis(glutathionyl)spermidine cofactor for glutathione reductase in trypanosomes. Science 227:1485.
2. Fairlamb AH, Cerami, A (1985). Identification of a novel, thiol containing cofactor essential for glutathione reductase

enzyme activity in trypanosomatids. Molec Biochem Parasitol 14:187.
3. Henderson GB, Fairlamb AH, Cerami A (1987). Trypanothione dependent peroxide metabolism in *C. fasciculata*. Mol Biochem Parasitol 24:39.
4. Meister A, Anderson, M (1983) Glutathione. Annu Rev Biochem 52:711.
5. Docampo R, Moreno SNJ. (1984) Free radical intermediates in the antiparasitic action of drugs and phagocytic cells. in Free Radicals in Biology vol. 6 pp 243, Academic Press.
6. Shames SL, Fairlamb AH, Cerami A, Walsh CT (1986). Purification and characterization of trypanothione reductase from *C. fasciculata*, a newly discovered member of the family of disulfide containing flavoprotein reductases. Biochemistry 25:3519.
7. Krauth-Siegel RL, Enders B, Henderson GB, Fairlamb AH, Schirmer RH. (1987). Trypanothione reductase from *T. cruzi*, purification and characterization of the crystalline enzyme. Eur J Biochem 27:123.
8. Shames SL, Kimmel BE, Peoples OP, Agabian N, Walsh CT (1988). Trypanothione reductase of T. congolense: gene isolation, primary sequence determination, and comparison to glutathione reductase. Biochemistry 27:5014.
9. Fox B, Walsh CT (1982). Purification and characterization of a transposon encoded flavoprotein containing an oxidation reduction active disulfide. J Biol Chem 257:2498.
10. Williams CH (1975). in The Enzymes (Boyer, PD ed) 3rd Ed vol 13, pp 89.
11. Krauth-Siegel RL, Blatterspiel R, Saleh M, Schiltz E, Schirmer RH, Untucht-Grau R (1982) Glutathione reductase from human erythocytes. The sequences of the NADPH domain and of the Interface domain. Eur J Biochem 121:259
12. Pai EF, Schulz GE (1983). The catalytic mechanism of glutathione reductase as

derived from x-ray diffraction analysis of reaction intermediates. J Biol Chem 258:1752.

13. Karplus PA, Pai EF, Schulz GE (1989). A cystallographic study of the glutathione binding site of glutathione reductase at 0.3 nm resolution. Eur. J. Biochem., in press.
14. Karplus PA, Schulz GE (1987). Refined structure of glutathione reductase at 1.5 A resolution. J Molec Biol 195:701.
15. Miller SM, Moore, MJ, Massey V, Williams CH Jr, Distefano MD, Ballou DP, Walsh CT (1989) Evidence for the participation of cys 558 and cys 559 at the active site of mercuric reductase. Biochemistry 28:(in press).
16. Moore MJ, Walsh CT (1989). Mutagenesis of the N- and C- terminal cysteine pairs of Tn501 mercuric ino reductase: consequences for bacterial detoxification of mercurials. Biochemistry 28:(in press).
17. Henderson GB, Fairlamb AH, Ulrich P, Cerami A (1987). Substrate specificity of the flavoprotein trypanothione disulfide reductase from *C. fasciculata* Biochemistry 26:3023.
18. Greer S, Perham RN (1986). Glutathione reductase from *E. coli*, cloning and sequence analysis fo the gene and relationship to other flavoprotein disulfide oxidoreductases. Biochemistry 25:2736.
19. Tabor S, Richardson CC (1985). A bacteriophage T7 RNA polymerase/promoter system for controlled exclusive expression of specific genes. Proc Natl Acad Sci 82:1074.
20. Fox B, Walsh CT (1983). Mercuric reductase: homology to glutathione reductase and lipoamide dehydrogenase. Iodoacetamide alkylation and sequence of the active site peptide. Biochemistry 22:4082.

Protein and Pharmaceutical
Engineering, pages 135–144

ENZYME MIMICS [1]

Ronald Breslow

Department of Chemistry, Columbia University
New York, New York 10027

ABSTRACT Enzyme mimics can help us understand enzymes themselves, and they can also perform useful chemical transformations with enzyme-like rates and selectivities. Examples of both situations are described.

INTRODUCTION

Enzyme models and mimics have two functions, which differ in the direction of information flow. Some mimics take chemical information and use it to help us understand biochemical processes. We understand things by relating them to other facts that we understand better, so the understanding of enzymatic reactions naturally involves their explanation in terms of 'simple" organic and inorganic chemistry. Since the needed simple chemistry does not always pre-exist, one function of enzyme mimics is to generate it.

The second function takes information from biochemistry and applies it to organic chemistry. Natural catalysts are remarkably effective at performing rapid reactions with great selectivity. They select the **substrate**, the **reaction** to be performed, the **region** of the substrate in which the reaction will occur, and the **stereochemistry** of the process.

---

[1]This work was supported by the NIH, NSF, and ONR.

Biomimetic chemistry is the branch of organic chemistry that takes its inspiration from biochemistry, and that produces artificial enzymes to mimic enzymes, either for purely scientific reasons or with practical transformations in mind (1).

In this symposium I will describe examples of each situation, to illustrate the current state of our work in this field.

MIMICS OF RIBONUCLEASE A

The enzyme ribonuclease A performs the cleavage of RNA in a two-step process. First the phosphate group is attacked by the 2' OH of a ribose, expelling the 5' oxygen of the next nucleoside unit and forming a 2',3' cyclic phosphate. Then the cyclic phosphate is hydrolyzed by the enzyme, to form a 3' phosphate as a terminal residue. Both steps are catalyzed by the same enzyme functional groups, the imidazole rings of His-12 and His-119 along with the ammonium cation of Lys-41. The catalysis shows a bell-shaped pH *vs.* rate profile, indicating that one histidine is present as the imidazole (Im) while the other acts in the protonated imidazolium (ImH^+) form. The mechanism usually written involves the Im acting as a base while the ImH^+ protonates the leaving group (2).

We have been developing mimics for ribonuclease, and in the course of this work we have investigated the ability of simple imidazole units, in imidazole buffer, to catalyze the cleavage of RNA (3-6). We find that indeed they do catalyze the reaction, but by a mechanism different from that usually considered for ribonuclease. We have elucidated this mechanism for the enzyme mimic, and have shown that it is also the preferred mechanism for an artificial enzyme that uses the enzyme's catalytic groups.

An examination of the data on the enzyme itself suggests that our novel mechanism is probably also used by the enzyme. The ImH^+ group first protonates the phosphate anion, then the basic Im promotes attack by the 2' OH. Later the phosphorane intermediate loses the extra proton again, and it is transferred to the leaving group.

This mimic has thus given us additional insight into the operation of the enzyme itself, and it has also laid the foundation for improved synthetic catalysts.

TRANSAMINASE MIMICS

The metabolism of amino acids is dominated by enzymes that utilize pyridoxal phosphate and pyridoxamine phosphate. Transamination involves first the reaction of pyridoxal phosphate with an amino acid to produce pyridoxamine phosphate and a ketoacid. Then the pyridoxamine reacts with a different ketoacid to produce a new amino acid with regeneration of the pyridoxal phosphate. Other transformations of amino acids also use pyridoxal phosphate as a coenzyme.

We have pursued a program aimed at building artificial enzymes that incorporate the key features of such enzymatic processes. We build in a binding group to select a particular substrate and to promote the reaction. We incorporate a pyridoxal or pyridoxamine unit to catalyze or perform the transformations. We also build in a basic catalytic group to perform the proton transfers required. Finally, we see to it that the basic group is chirally placed so that it can direct the formation of optically active amino acids (7-15).

We have described compounds in which pyridoxamine is covalently attached to the primary face of β-cyclodextrin (7), and to the secondary face (8). In both cases the result was an artificial enzyme that binds hydrophobic substrates such as phenylpyruvic acid preferentially, and then produces the derived amino acid-- in this case phenylalanine. Furthermore, some turnover was seen in reactions involving this ketoacid with another amino acid, so we are looking at true catalysis. A related artificial enzyme, combining the coenzyme with a binding group, was produced with pyridoxamine and a synthetic hydrophobic cavity molecule (9).

In the enzymes there are proton transfers catalyzed and directed by basic groups, so we incorporated these as well. First we simply established that such basic groups do indeed perform the proton transfers that interconvert ketoacid derivatives with amino acid derivatives (10); then we showed that when the basic group is placed on an asymmetric center, so it can reach only one face of the transamination intermediate, it does indeed produce optically active amino acids (11). Then we combined the two lines.

In our own laboratory we synthesized compound 1, and saw that it indeed acted as a synthetic transaminase (14). It selected aromatic ketoacids that could bind into the cavity and selectively converted them to amino acids. It accelerated the reaction with the basic group, and also used it to direct the formation of optically active amino acids. However, the chiral preference was not large, only 7:1.

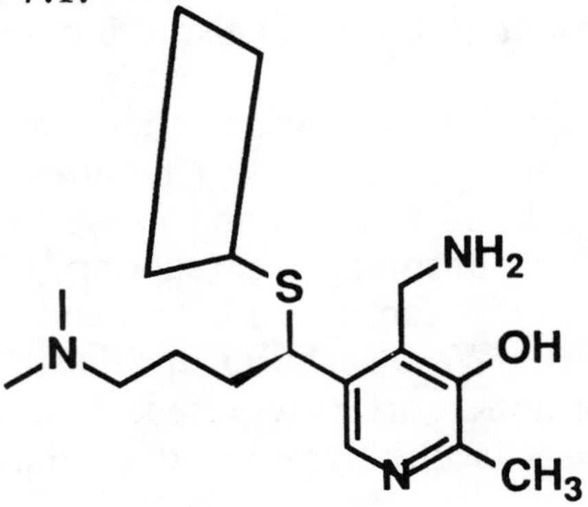

1

In the Tabushi lab, in a loose collaboration with us, a different approach to these compounds was examined. Compound 2 was prepared, and it showed excellent optical induction (15).

In our recent work we have improved these systems quite a bit. We have prepared compound 3 and compound 4. In both of these cases several degrees of freedom have been frozen out, so we expect that the cooperation between binding group and catalytic group will be much more effective. The compounds are still being evaluated.

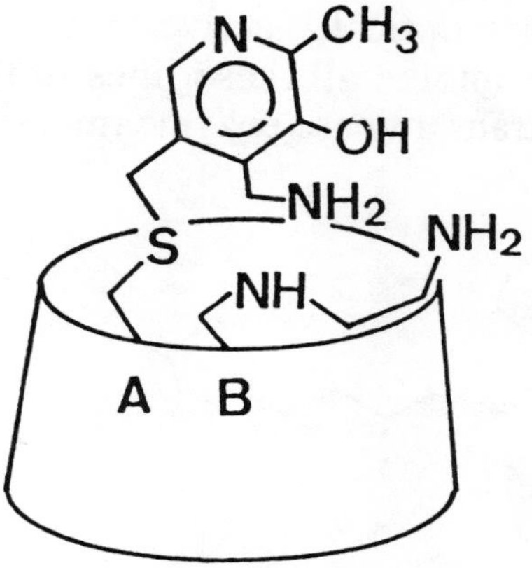

N CH3
OH
NH2
S
NH2
NH
A B

2

CHO
OH
S S
N
CH3

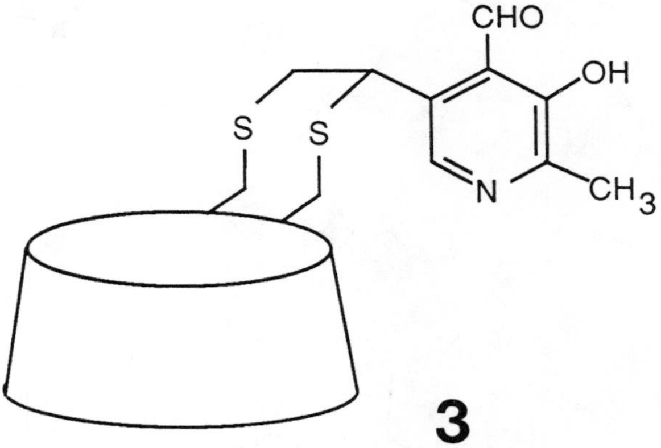

3

We have found that an ethylenediamine unit is particularly effective in catalyzing the needed proton transfers, since one nitrogen can remove the proton and the other can attach it at a more remote position (14). We have attached such a unit to a bridged pyridoxamine in compounds **5** and **6**, and find that the transamination is very fast. Optically active versions, being prepared, should be quite interesting.

The ethylenediamine unit is also incorporated in a compound related to **1**, which we have prepared in optically active form.

Again the proton transfer step of transamination is very rapid, and the products are optically active amino acids. In future work we hope to combine all these lines in the production of very fast effective transaminase enzyme mimics.

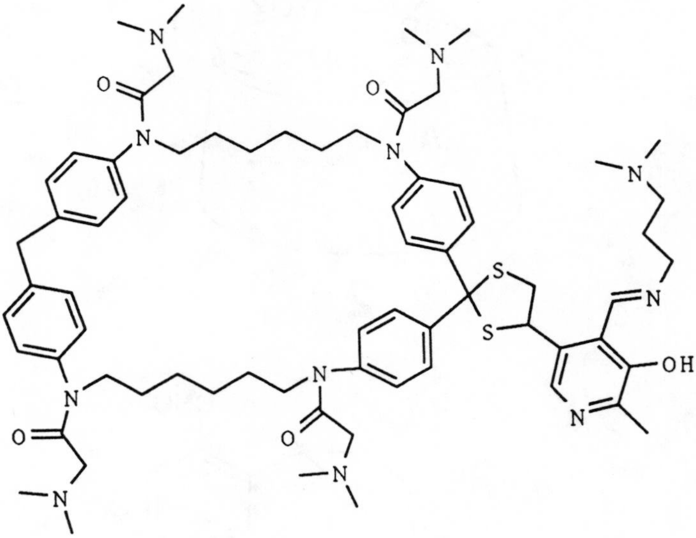

N
O
N
N
O
N
N
S
S
N
OH
N
O
N
N
O
N
N

4

H_2N
HO
S
N
S
H
N
N
H

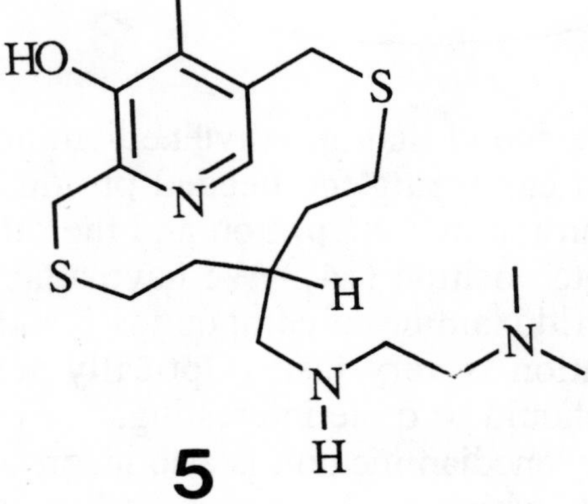

5

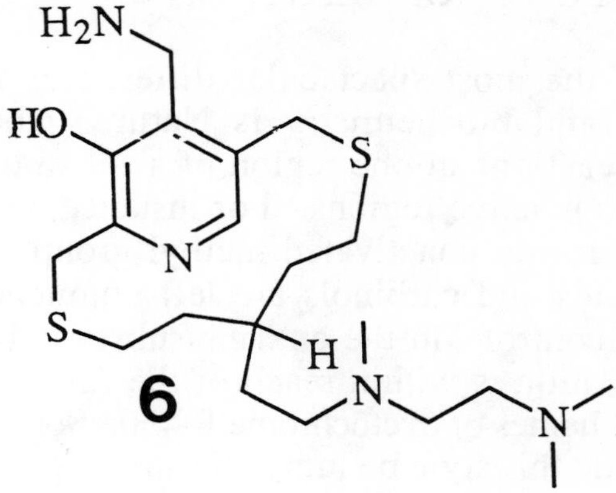

6

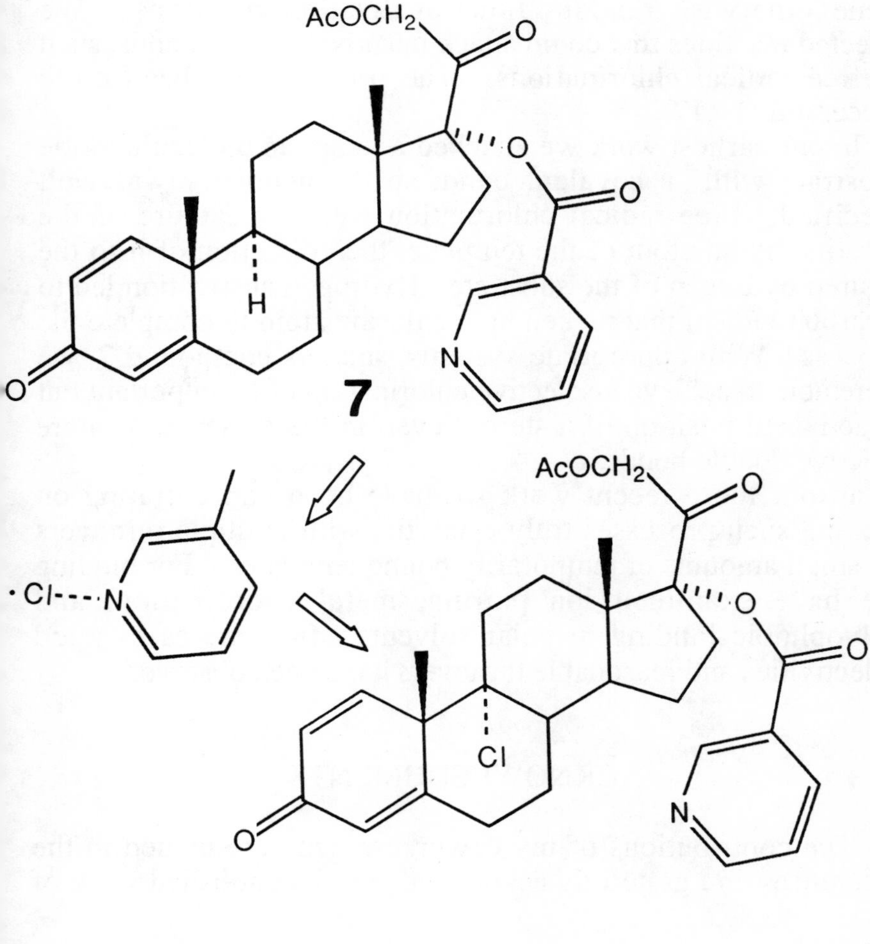

7

REMOTE FUNCTIONALIZATION

One of the most spectacular differences between organic chemistry and biochemistry is Nature's ability to perform selective reactions in one region of a substrate while ignoring much more reactive regions. For instance, in the biosynthesis of cholesterol an unactivated methyl group is oxidized while double bonds and carbinols are left untouched. The trick is geometric control: in the enzyme/substrate complex only the desired position is within reach of the functionalizing groups, such as the hemes of cyctochrome P-450. Some years ago we set out to mimic this style be imposing strong geometric control on some otherwise random functionalization reactions. We selected reactions that could attack unactivated C-H bonds, such as free radical chlorinations. The program has been quite successful (16,17).

In our earliest work we attached a template molecule to the substrate with a covalent bond, so the geometry was well specified. Free radical chlorination went by capture of the chlorine by an atom of the template, then direction of it to the desired hydrogen of the substrate. Hydrogen abstraction led to a carbon radical that picked up a chlorine atom to complete the process. With appropriate systems, such as compound 7, we were able to achieve a selective chlorination of an important but unactivated position of a steroid even in the presence of more reactive double bonds.

In our more recent work we have been concentrating on making such processes truly catalytic, with multiple turnovers by small amounts of temporarily bound templates. For binding we have examined ion pairing, metal coordination, and solvophobic binding in polar solvents. In some cases good selectivities and reasonable turnovers have been observed.

ACKNOWLEDGMENTS

The contributions of my coworkers, who are named in the references, are gratefully acknowledged. Unpublished work of

Mr. T. Liu, Dr. J. Canary, Dr. M. Sprecher, Dr. M. Kotera, Dr. B. Jursic,and Dr. S. T. Waddell is included in this article.

REFERENCES

1. Breslow R (1972). Biomimetic Chemistry. Chem Soc Rev 1:553.
2. Richards FM, Wycoff HW (1971). Ribonuclease A. In Boyer H: "The Enzymes" vol 4:647.
3. Corcoran R, LaBelle M, Czarnik AW, Breslow R (1985). An assay to determine the kinetics of RNA cleavage. Anal Biochem 144:563.
4. Breslow R, LaBelle M (1986). Sequential general base-acid catalysis in the hydrolysis of RNA by imidazole. J Am Chem Soc 108:2655.
5. Anslyn E, Breslow R (1989). On the mechanism of catalysis by ribonuclease. Cleavage and isomerization of the dinucleotide UpU catalyzed by imidazole buffers. J Am Chem Soc 111: in press.
6. Breslow R, Huang D-L, Anslyn E (1989). On the mechanism of action of ribonucleases: dinucleotide cleavage catalyzed by imidazole and Zn^{2+}. Proc Nat Acad Sci USA : in press.
7. Breslow R, Hammond M, Lauer M (1980). Selective transamination and optical induction by a beta-cyclodextrin-pyridoxamine artificial enzyme. J Am Chem Soc 102: 421.
8. Breslow R, Czarnik AW (1983). Transaminations by pyridoxamine selectively attached at C-3 in beta-cyclodextrin. J Am Chem Soc 105:1390.
9. Winkler J, Coutouli-Argyropoulou E, Leppkes R, Breslow R (1983). Artificial transaminase carrying a synthetic macrocyclic binding group. J Am Chem Soc 105:7198.
10. Zimmerman SC, Czarnik AW, Breslow R (1983). Intramolecular general base-acid catalysis in transaminations catalyzed by pyridoxamine enzyme analogues. J Am Chem Soc 105:1694.

11. Zimmerman SC, Breslow R (1984). Assymetric synthesis of amino acids by pyridoxamine enzyme analogues utilizing general base-acid catalysis. J Am Chem Soc 106:1490.
12. Weiner W, Winkler J, Zimmerman SC, Czarnik AW, Breslow R (1985). Mimics of tryptophan synthetase and of biochemical dehydroalanine formation. J Am Chem Soc 107:5544.
13. Breslow R, Czarnik AW, Lauer M, Leppkes R, Winkler J, Zimmerman S (1986). Mimics of transaminase enzymes. J Am Chem Soc 108:1969.
14. Breslow R, Chmielewski J, Foley D, Johnson B, Kumabe N, Varney M, Mehra R (1988). Optically active amino acid synthesis by artificial transaminase enzymes. Tetrahedron 44:551.
15. Tabushi I, Kuroda Y, Yamada M, Higashimura H, Breslow R (1985). A-(modified B-6)-B-[omega-amino(ethylamino)]-beta-cyclodextrin as an artificial B-6 enzyme for chiral aminotransfer reaction. J Am Chem Soc 107:5545.
16. Breslow R (1986). Artificial enzymes and enzyme models. Adv in Enzym 58:1.
17. Breslow R (1988). Biomimetic regioselective template-directed functionalizations. Chemtracts Org Chem 16:408.

Protein and Pharmaceutical Engineering, pages 145–158

PROBING THE ACTIVE SITE OF THE LEUCINE-BINDING PROTEINS OF *E. COLI*[1]

Dale L. Oxender, David J. Maguire[2], and Mark D. Adams[3]

Department of Biological Chemistry
University of Michigan
Ann Arbor, Michigan 48109

ABSTRACT High affinity branched-chain amino acid transport in *E. coli* involves two periplasmic binding proteins and three membrane components. The branched-chain amino acids diffuse through the outer membrane of *E. coli* and form a complex with the soluble binding proteins. These complexes interact with the inner membrane components to produce the transport of the ligand. Two periplasmic leucine-binding proteins have been isolated, purified, and crystallized. The three-dimensional structures have been determined to high resolution by Quiocho and co-workers (7; 8). The genes for these proteins *livJ* and *livK* have been cloned and sequenced. They encode the leucine-isoleucine-valine (LIV) and leucine-specific (LS) binding proteins, respectively. The amino acid sequences of LIV-BP and LS-BP are 78% homologous, suggesting that they have derived from a gene duplication event. The X-ray structural analysis by Quiocho suggests that six amino acid residues line the leucine binding pocket of LIV-BP. Three of these residues – leu_{77}, cys_{78}, and ala_{101} are conserved in the two proteins. The three amino acid residues that are not conserved are tyr_{18}->trp, ala_{100}->gly, and phe_{276}->trp. We have prepared hybrid

---

[1] This work was supported by NIH grant GM 11024.
[2] Permanent address: Division of Science and Technology, Griffith University, Brisbane, Australia, Q4111
[3] MDA was supported by a Horace H. Rackham pre-doctoral fellowship.

proteins by gene splicing techniques and specific mutants using site-directed mutagenesis in an attempt to understand the molecular basis of the difference in specificities of the LIV- and LS-binding proteins.

INTRODUCTION

The periplasmic binding proteins offer unique opportunities for the study of the first step in the transport process, namely binding, in isolation from the membrane translocation process. These proteins are components of high affinity transport systems for ions, carbohydrates, and amino acids. Some periplasmic binding proteins also participate in recognition for chemotaxis. In contrast with low affinity, high capacity transport systems which exhibit a K_D for transport of about 10^{-3}M, the high affinity systems exhibit a K_D of approximately 10^{-7}M and require phosphate bond hydrolysis to carry out transport of the substrate across the membrane. With the exception of maltodextrins, transport substrates diffuse across the outer membrane through non-specific porins; the first specific interaction is with either the single membrane component (low affinity systems) or one of the periplasmic binding proteins. For review see (1).

In addition to a common general role in transport, the high-affinity transport systems share other characteristics. At the genetic level, a similar operon organization is seen for many of the transport systems. Typically, a polycistronic transcript encodes a binding protein gene followed by the genes for the membrane components of the system. Occasionally the binding protein gene is on a separate transcript, but in all cases where the information is available, the binding protein gene is located 5' from the membrane component genes. Another common feature of these systems is that each binding protein appears to contain a single substrate binding site and to interact with the membrane components as a monomer.

A great deal has been contributed to the understanding of binding protein structures by Quiocho and co-workers, who have determined the X-ray crystal structure of members of each group of binding proteins: ion-binding, carbohydrate-binding, and amino acid-binding (2). Even though only very limited primary sequence homology is observed among the family of binding proteins, a significant amount of tertiary structure homology is present in the three-dimensional structures. The proteins have two globular domains, each consisting of a central, mostly parallel, ß-sheet flanked by two or three α-helices. The two domains are connected by three segments which are structurally in close proximity to each other and which form the base of a deep cleft between the two domains. The substrate binding pocket lies in this cleft, exposed to solvent molecules in the unbound form of the protein. Binding of substrate is presumed to be accompanied by a closing of the cleft, with the segments connecting the two domains acting as a hinge. This leads to the sequestering of the substrate from the solvent and presumably facilitates interaction of the complex with the membrane-bound components of the transport system.

Because of the overall similarity in three-dimensional structure among the group of periplasmic binding proteins, despite a large divergence in primary structure, binding proteins also constitute a unique model system for studying the rules for protein folding. It has been well established that the primary sequence is sufficient to determine the final three-dimensional shape of a protein. What remains to be elucidated is the exact way in which that information is carried and elaborated. The binding proteins, with their divergent sequence, yet common structural characteristics, should prove useful in examining these rules of folding (14; 15).

The two leucine-binding proteins of *E. coli* present an additional unique opportunity to focus on the relationships between structure and function. Both binding proteins interact with the same set of membrane-associated components for the branched-chain amino acid transport system LIV-I. The gene for the leucine specific-binding protein, *livK*, lies 5' to the genes for the membrane components in a polycistronic message. *LivJ*, which codes for the LIV-BP, is located adjacent to *livK* at minute 74 on the *E. coli* chromosome (3). The two proteins share about 78% sequence identity (Fig. 1), but have different substrate specificities (Table 1) (4; 5). LS-BP is specific for leucine while LIV-BP binds leucine, isoleucine, and valine with approximately equal affinity.

Although the leucine-binding proteins have both recently been crystallized in the non-liganded state, suitable crystals of the liganded binding proteins have not been obtained. In contrast, all other binding proteins which have been analyzed by X-ray crystallography were crystallized in the liganded form. However, diffusion of leucine into crystals of LIV-BP reveals a discrete site of primary interaction with the binding protein (7; 8). Leucine is bound solely to the N domain on the inner face of the cleft, in contrast to the binding sites known for other binding proteins, where both N and C domains form hydrogen bonds with the substrate. While none of the

```
         10        20        30        40        50        60        70        80
          |         |         |         |         |         |         |         |
DDIKVAVVGAMSGPIAQWGIMEFNGAEQAIKDINAKGGIKGDKLVGVEYDDACDPKQAVAVANKIVNDGIKYVIGHLCSS
 ************* ** *  ** *****  ********** **    **************** ***************
EDIKVAVVGAMSGPVAQYGDQEFTGAEQAVADINAKGGIKGNKLQIAKYDDACDPKQAVAVANKVVNDGIKYVIGHLCSS

         90       100       110       120       130       140       150       160
          |         |         |         |@@       |         |         |         |
STQPASDIYEDEGILMISPGATAPELTQRGYQHIMRTAGLDSSQGPTAAKYILETVKPQRIAIVHDKQQYGEGLARSVQD
**************** * ******* **** * ** **** *********** ******************** ***
STQPASDIYEDEGILMITPAATAPELTARGYQLILRTTGLDSDQGPTAAKYILEKVKPQRIAIVHDKQQYGEGLARAVQD

        170       180       190       200       210       220       230       240
          |         |         |         |         |         |         |         |
GLKAANANVVFFDGITAGEKDFSALIARLKKENIDFVYYGGYYPEMGQMLRQARSVGLKTQFMGPEGVGNASLSNIAGDA
***  ***************** * *************** ***** *****  ************ * *******
GLKKGNANVVFFDGITAGEKDFSTLVARLKKENIDFVYYGGYHPEMGQILRQARAAGLKTQFMGPEGVANVSLSNIAGES

        250       260       270       280       290       300       310       320
          |         |         |         |         |         |         |         |
AEGMLVTMPKRYDQDPANQGIVDALKADKKDPSGPYVWITYAAVQSLATALERTGSDEPLALVKDLKANGANTVIGPLNW
*** *** ** *** ***  **** ** * ****  ** **** ***  *      * *    * ****   ** *** *
AEGLLVTKPKNYDQVPANKPIVDAIKAKKQDPSGAFVWTTYAALQSLQ AGLNQS DDPAEIAKYLKANSVDTVMGPLTW

        330       340
          |         |
DEKGDLKGFDFGVFQWHADGSSTAAK
********* **** *** *  * **
DEKGDLKGFEFGVFDWHANGTATDAK
```

Notes:
* = Sequence identity
@@ = Position of StyI site in coding sequence
<u>X</u> = Residues in leucine binding site (LIV-BP)
X = Divergent residues in side chain binding site

FIGURE 1. Primary sequences of LS-BP (top) and LIV-BP (bottom).

conformational changes assumed to occur upon ligand binding occur in the crystal, this work nevertheless provides a first approximation of the leucine binding site and opens the possibility of examining the determinants of specificity on a residue by residue basis.

METHODS

Genes for LS-BP and LIV-BP have been cloned and sequenced (4; 6). Single crystal X-ray diffraction structures were calculated by Sack and co-workers (7; 8). Molecular modelling was carried out using MOGLI on an Evans and Sutherland PS390 graphics workstation. Site-directed mutageneses to introduce common restriction sites and to alter residues at the putative ligand binding site were performed as described by Su and El-Gewely(9), a modification of the Kunkel method(10). Transport assays were performed as in (11). Leucine or proline uptake was assayed for 15 sec at 37°C using 0.1 μM radioactive amino acid with a specific activity of 5.5 Ci/mmol (^{3}H-leucine) or 295 mCi/mmol (^{14}C-proline). For inhibitor studies, the analog was present at either 5 or 50 μM. Transport measurements were made in *E. coli* strain AE840218 (*malA1 xyl-7 mtl-2 argG6 his-1 trp-31 str-104 nal pdxC3 livR livJ*). Western blots were performed according to (12).

TABLE 1.
LEUCINE-BINDING SPECIFICITIES OF PURIFIED LIV- AND LS-BP[a]

Inhibitor	LIV-BP	LS-BP
Isoleucine	10	97
Trifluoroleucine (TFL)	95	15

[a]Binding is expressed as percent of uninhibited activity at 200 μM inhibitor concentration and 2.5 μM ^{14}C-L-leucine (13).

RESULTS

The X-ray crystal structures of the LS-BP and LIV-BP are remarkably similar (8). Therefore the ligand binding site is also expected to be very similar for both proteins. Among the residues identified by Sack *et al* (7) as contacting the leucine side chain, three are conserved (leu_{77}, cys_{78}, and ser_{79}) and three are divergent between the two proteins (Fig 1). All six residues are located in the N domain. As part of our study of the mechanism of leucine transport, we wanted to study the potential role of these residues, if any, in binding and/or transport.

We have taken two approaches to this problem. The first is to make site-specific changes to the residues mentioned above to probe the role of each residue in leucine binding and the discrimination among the transport substrates. Second, if the putative site is substantially correct, then large portions of the C domain are not necessary for binding (although they may still play a role in transport); switching the C domain of LS-BP with the C-domain of LIV-BP (and vice versa) should not affect binding. These mutations are presented in schematic form in Figure 2.

Switching the domains to create hybrid binding proteins requires the presence of a common restriction

		18	77	78	100	101		276		
wild type		tyr	leu	cys	ala	ala		phe	(LIV)	
		trp	leu	cys	gly	ala		tyr	(LS)	
MUTANTS	JSty	Y	L	C	A	A	*	F		SITE-DIRECTED
	KSty	W	L	C	G	A	*	Y		
	JK	Y	L	C	A	A	*	Y		GENE FUSION
	KJ	W	L	C	G	A	*	F		
	JAG100	Y	L	C	G	A	*	F		SITE-DIRECTED
	KGA100	W	L	C	A	A	*	Y		

* = StyI restriction site

FIGURE 2. Generation of domain-switched and site-directed mutants.

site in the two genes. Despite the high degree of homology, there are no conserved restriction sites in the genes for LS- and LIV-BP. Ideally, the position of such a site would cleanly separate the two domains, but since in fact each domain is composed of residues from both the amino and carboxy terminal half of the polypeptide chain, this is impractical. Residue 120 was chosen as the best candidate for the position of the switch since it separates the bulk of the N domain from the bulk of the C domain, preserves most of the ligand binding site in the N domain, and is amenable to generation of a silent restriction site.

A unique StyI restriction enzyme cleavage site was engineered by site-directed mutagenesis into the LS-BP (pOX7) and LIV-BP (pOX15) genes at the position encoding amino acid residues 120-121 in such a way that the amino acid sequence was unchanged. The two mutant plasmids created are referred to as pKSty and pJSty respectively.

These two plasmids were used to create mutant binding protein genes. First, domain-switch or hybrid genes were constructed by *in vitro* recombination of *livJ* and *livK* plasmids each containing an engineered StyI site, resulting in plasmids pJK and pKJ.

Second, using site-directed mutagenesis, two mutant plasmids (pKGA100 and pJAG100) were constructed from pKSty and pJSty. pJAG100 contains *livJ* with an ala->gly change at codon 100. pKGA100 contains *livK* with gly->ala at codon 100.

High-affinity leucine transport was measured by assaying uptake of 0.1 μM ^{3}H-leucine. Inhibitors of transport were used to determine the specificity characteristics of the mutant proteins. These results are shown in Figure 3.

From the limited number of experiments performed to date, it appears that the domain-switched proteins show a decreased leucine transport activity and moderate sensitivity to both isoleucine (ile) and trifluoroleucine (TFL) (Fig. 3a). Western blot analysis shows that the domain-switched proteins are expressed at wild-type levels (not shown). Cultures of cells containing the plasmid pJAG100 appear to transport leucine at or about the same activity as the wild type level (Fig. 3a). There seems to be some loss of sensitivity to isoleucine inhibition, and some increase in sensitivity to TFL inhibition (Table 2). Interpretation of the inhibition pattern of pKGA100 was not possible due to the presence of a chromosomally-encoded wild-type LivK protein in the strain AE8401218.

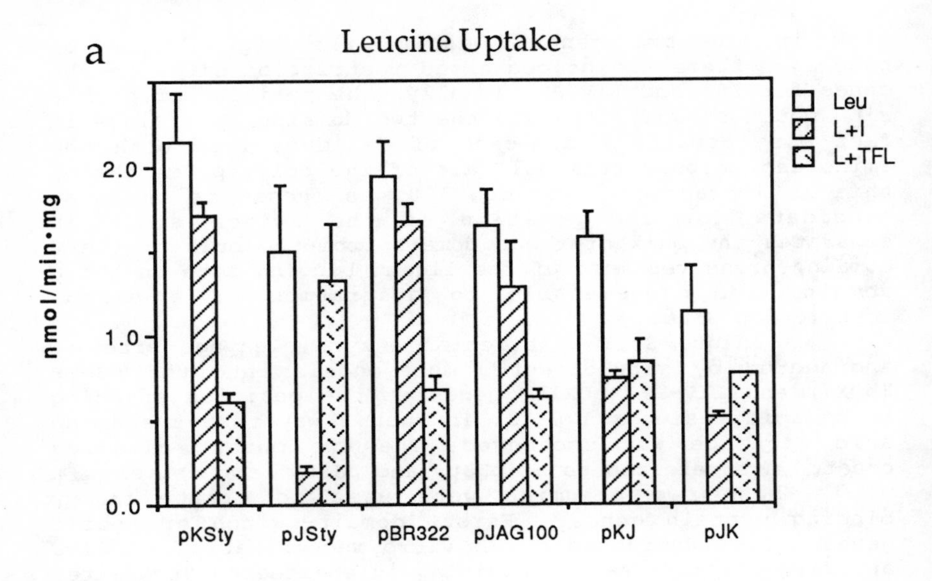

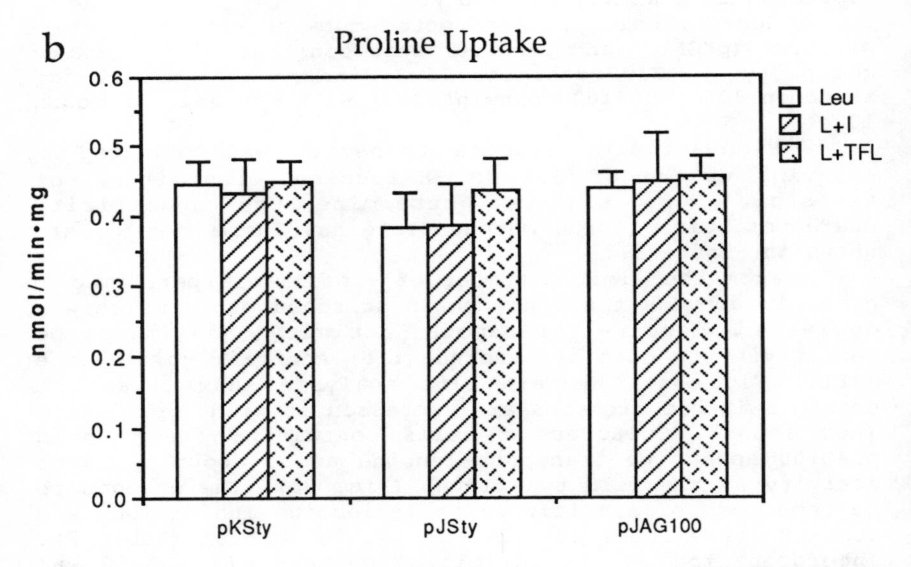

FIGURE 3. Uptake of ^{3}H-leucine (a) and ^{14}C-proline (b) in the presence and absence of transport inhibitors. Bars represent the mean of three independent cultures.

TABLE 2.
PERCENT INHIBITION OF LEUCINE TRANSPORT

% of uninhibited transport		
Plasmid	+Ile	+TFL
pKSty	80	29
pJSty	13	88
pBR322	87	35
pJAG100	78	38
pKJ	45	68
pJK	47	53

[a]Inhibition patterns are from Figure 3a.

None of the amino acids or analogs affecting the LIV system are transported by the proline transport system. Therefore, proline transport should not be affected by the differences in binding protein nor by the presence of LIV system inhibitors. Figure 3b shows ^{14}C-proline transport measured concurrently with ^{3}H-leucine transport. Proline transport is essentially unchanged in all clones tested under all experimental conditions.

DISCUSSION

Quiocho's group has identified the amino acid residues responsible for the primary binding of leucine to crystallized LIV-binding protein. Of the six residues which make direct contact with the leucine side chain, three are conserved in LS-BP. Moreover, all residues proposed to contribute H-bonds to the α-amino and α-carboxyl moieties of leucine are conserved between the two proteins. Analysis of this binding site and the construction and testing of proteins mutated in this region should provide us with information about the relationship between molecular architecture and binding activity in these binding proteins.

Although the two proteins have 78% primary sequence identity and the crystals show an overall root-mean-squared difference of 0.68Å of the α-carbon positions of the equivalent residues, there are nevertheless significant sequence differences between the two proteins, including a difference in overall sequence length of two residues (Fig. 1). Careful scrutiny of stereoscopic diagrams in which the structures of the two molecules have

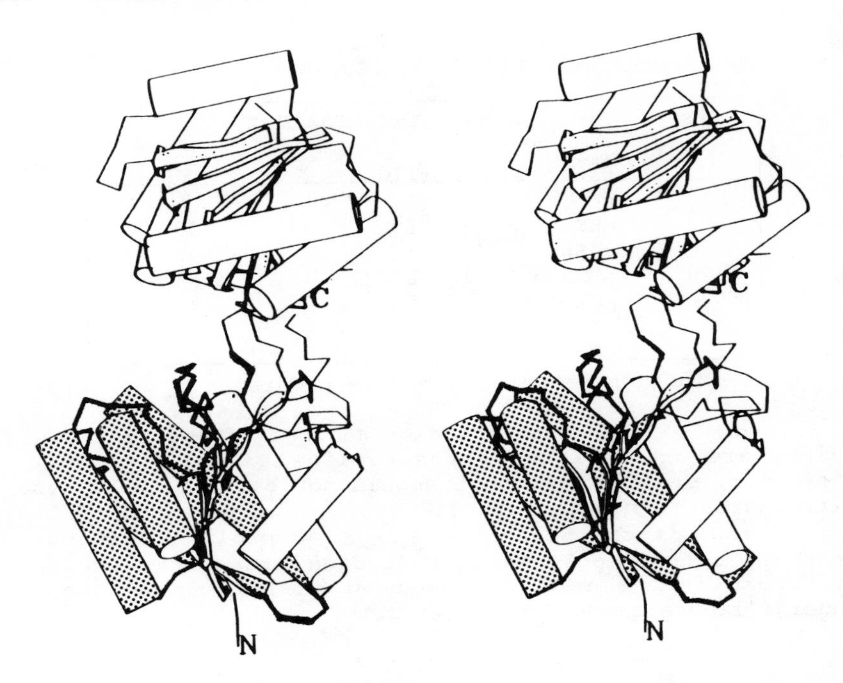

FIGURE 4. Stereo-view of model of LIV-BP showing extent of domain switching. Shaded portion includes N-terminal residues up to 120.

been superimposed reveals that areas of the greatest tertiary structure divergence are located on surface loops. The deletion of two residues is located in the loop between helix XI and strand K in the N domain. The other region where the two backbone structures diverge significantly is in the C-domain at residues 163-166, located on a loop between helix V and strand G. Both of these regions of major divergence in tertiary structure are far-removed from the primary binding site identified by Sack, *et al.* (8).

Since the crystal structures show such similar overall shape and domain architecture, differences in specificity between the two proteins must be attributable to non-homologous residues within the respective binding clefts. These differences could accompany a major shape change induced by an alteration in the proximity of the two sides of the binding cleft such as would occur during operation of the "hinge closure" mechanism proposed by Quiocho (16). The alterations in protein tertiary structure created by the domain switch experiments reported here might be expected to affect interactions between domains during closure of the cleft. It is also conceivable, however that specificity is related to discrete differences within the binding site identified by Quiocho, and which therefore must be attributed to primary sequence differences in that region. The point mutations created in the work reported here reflect the smallest change that can conceivably be made between two amino acid side-chains within that site, namely interchanging a methyl group and a hydrogen. As pointed out by Sack, such changes *per se* would not be expected to cause the sort of steric hindrance required to confer specificity upon the LS binding protein (8). The data presented here are consistent with the suggestion that the ala_{100}->gly mutation created in the proposed site affects the binding characteristics of LIV-BP. This supports the proposal that this site is the true primary binding site. It should be noted that the inhibition by TFL may not be due to interaction in the same manner as isoleucine; we have not yet tested for competition between the two inhibitors.

The strain used for the measurement of transport activity (AE840218) is a *livJ* mutant, but does contain a *livK* gene. The same transport activity is observed when pBR322 or pKSty (*livK* containing plasmid) is present in the cell. Nonetheless, it is possible to distinguish the predominating form of binding protein participating in transport through the use of inhibitors. Isoleucine competes only with LIV-BP and trifluoroleucine (TFL) inhibits mainly LS-BP (Table 1). Thus the patterns of inhibition of transport of KSty and JSty are very different, even though the strain background is $livK^+$ in each case. The pattern of inhibition is representative of the plasmid encoded gene, not the chromosomally encoded gene. However it is still possible that the production of an inactive protein, or no protein, from pJAG100 might also be an explanation for the observed results.

Since there is no X-ray crystallographic data available for the mutant proteins, it is impossible to say whether a deviation from wild type structures has occurred at the primary site of binding. A stereoscopic model of LIV-BP is presented in Fig. 4 in which the N-terminal residues up to 120 have been highlighted in order to show what proportion of the N-domain is involved in the switch. Provided there is no major change in relative positions of residues within the binding pocket, the only change in the residues of the leucine binding site would be to interconvert tyrosine and phenylalanine at position 276. However, it is possible that the observed changes in transport activity of each domain-switched mutant protein are due to an alteration in the overall shape of the molecule, thereby interfering with the interaction of the binding protein with membrane components of the transport system. This could include poor intramolecular contacts leading to interferences with the "hinge closure" mechanism. It is also conceivable that several of these factors may play a role in determining the final activity observed.

The preliminary results obtained with $alanine_{100}$ to glycine point-mutants adds support to the use of a similar binding site by both proteins. Many of the arguments which were invoked when considering the effects of domain switching could also apply to the results obtained using point mutants. However, the more discrete nature of these latter mutations would be expected to result in considerably less disruption to the tertiary structure of resulting proteins. A molecular explanation for the alteration in specificity of binding for amino acids is not currently available and will likely require structural analysis of the mutant proteins. Future work will concentrate on the effects of mutating other binding site residues which are diverged between the two binding proteins.

Resolution of the uncertainties which we have described will have to await further characterization in a strain deficient in both wild type binding activities. Subsequent protein purification and binding studies will enable us to choose mutant proteins for further studies at the X-ray structure level.

ACKNOWLEDGEMENTS

We wish to thank F. Quiocho, for supplying us with the X-ray coordinates of the LIV-BP, and also for permission to reproduce parts of Fig. 4.

REFERENCES

1. Ames, G F-L (1986). Bacterial periplasmic transport systems: Structure, mechanism, and evolution. Ann Rev Biochem 55:397.
2. Quiocho, FA (1986). Carbohydrate-binding proteins: Tertiary structures and protein-sugar interactions. Ann Rev Biochem 55:287.
3. Anderson, JJ, Oxender, DL (1977). *Escherichia coli* transport mutants lacking binding protein and other components of the branched-chain amino acid transport systems. J Bacteriol 130:384.
4. Landick, R, Oxender, DL (1985). The complete nucleotide sequences of the *Escherichia coli* LIV-BP and LS-BP genes implications for the mechanism of high-affinity branched-chain amino-acid transport. J Biol Chem 260:8257.
5. Rahmanian, M, Claus, DR, Oxender, DL (1973). Multiplicity of leucine transport systems in *Escherichia coli* K-12. J Bacteriol 116:1258.
6. Oxender, DL, Anderson, JJ, Daniels, CJ, Landick, R, Gundalus, RP, Yanofsky, C (1980). Amino terminal sequence and processing of the precursor of the leucine specific binding protein and evidence for conformational differences between precursor and mature forms. Proc Natl Acad Sci USA 77:2005.
7. Sack, JS, Saper, MA, Quiocho, FA (1989). Periplasmic binding protein structure: Refined X-ray structures of the leucine/ isoleucine/valine-binding protein and its complex with leucine. J Mol Biol.206:171.
8. Sack, JS, Trakhanov, DS, Tsigannik, IH, Quiocho, FA (1989). Structure of the L-leucine-binding protein refined at 2.4Å resolution and comparison with the leu/ile/val-binding protein structure. J Mol Biol.206:193.
9. Su, T-Z, El-Gewely, MR (1988). A multisite-directed mutagenesis using T7 DNA polymerase: application for reconstructing a mammalian gene. Gene 69:81.
10. Kunkel, TA (1985). Rapid and efficient site-specific mutagenesis without phenotypic selection. Proc Natl Acad Sci 82:488.
11. Anderson, JJ, Quay, SC, Oxender, DL (1976). Mapping of two loci affecting the regulation of branched-chain amino acid transport in *Escherichia coli* K-12. J Bacteriol 126:80.
12. Davis, LG, Dibner, MD, Battey, JF (1986). "Basic Methods in Molecular Biology." New York: Elsevier, p 311.

13. Rahmanian, M, Claus, DR, Oxender, DL (1973). Multiplicity of leucine transport systems in *Escherichia coli* K-12. J Bacteriol 116:1258.
14. Gribskov, M, McLachlan, AD, and Eisenberg, D (1987). Profile analysis: Detection of distantly related proteins. Proc Natl Acad Sci 84:4355.
15. Gribskov, M, Homvak, M, Edenfield, J, and Eisenberg, D (1988). Profile scanning for three-dimensional structural patterns in protein sequences. Comput Appl Biosci 4:61.
16. Mao, B, Pear, MR, McCammon, JA, and Quiocho, FA (1982). Hinge-bending in L-arabinose-binding protein: The Venus's-flytrap model. J Biol Chem 257:1131.

Protein and Pharmaceutical Engineering, pages 159–166

MANIPULATIVE MUTAGENESIS OF ENZYMES[1]

Stephen C. Blacklow, David L. Pompliano, and Jeremy R. Knowles*

Department of Chemistry, Harvard University, 12 Oxford Street, Cambridge, Massachusetts 02138

ABSTRACT The value of making both site-directed changes and of generating random alterations in the amino acids of a protein molecule has been evaluated using the glycolytic enzyme, triosephosphate isomerase. The kinetic consequences of two site directed changes are described, and the results of a search for second site suppressor mutations across the whole gene of a sluggish mutant isomerase are discussed.

INTRODUCTION

There are two different approaches to the engineering of proteins using the powerful methods of molecular biology. In the first of these, the presumed importance of a particular amino acid, of a group of amino acids, or of an entire protein domain, is probed by changing or deleting it and examining the functional consequences. This approach is fruitful, and can often provide experimental evidence for what are otherwise only notions about function and mechanism. In its more ambitious forms, this line of attack has led to efforts to use site-specific mutagenesis in a rational way, for example to produce enzymes of altered specificity or repressors

---

[1]This work was supported by the National Institutes of Health, and Merck, Sharp and Dohme. The manuscript is taken largely from an extended abstract by the same authors, prepared for a conference on Prospects in Protein Engineering, in Gröningen, 1989.

that recognize a different operator sequence. Such work has produced mixed results, however, and it is clear that our current understanding of structure:relationships in proteins does not yet guarantee that rationally designed changes will always yield the predicted outcomes (1).

The second approach, in contrast, does not test our current views of the functional consequence of a particular structural change, but aims both to produce proteins having the desired characteristics, and to gain a deeper understanding of the problem by identifying all those structural changes that produce a particular functional effect. Random mutagenesis is used to produce a library of amino acid changes in a given protein, and a selection then identifies the mutants that confer the desired phenotype. Since this approach avoids all preconceived ideas about what is important, and since (at least in principle) a large fraction of mutant sequence 'space' can be searched, the potential exists for the unbiased discovery of alternative structural solutions to a given functional need. In a sense, of course, this random mutagenesis approach mimics the evolutionary refinement of biological function at the molecular level.

In the work described, both of these approaches (that is, the site-directed and the random mutagenesis methods) have been applied to one enzyme. The enzyme, triosephosphate isomerase, catalyzes the interconversion of the two triosephosphates, dihydroxyacetone phosphate and D-glyceraldehyde 3-phosphate (Figure 1) and two features of the enzyme's kinetic behavior show that it is a kinetically optimal catalyst for the levels of substrates that it experiences in muscle (2). First, in the thermodynamically downhill direction (with glyceraldehyde phosphate as substrate) the reaction is diffusion controlled. The highest transition state for the enzyme-catalyzed reaction is that involved in the binding of glyceraldehyde phosphate to the enzyme. Secondly, in the thermodynamically uphill direction with dihydroxy-acetone phosphate as substrate, and at the substrate levels known to obtain *in vivo*, the reaction is second order. The lowest ground state for the enzyme-catalyzed reaction is that of free enzyme plus free dihydroxyacetone phosphate. These two features allow us to conclude that triosephosphate isomerase could not improve as a catalyst. While our original work was done on the isomerase from chicken muscle, which from chemical and crystallographic studies appears to use the carboxylate of Glu-165 as a base and the imidazole of His-95 as an acid, every other isomerase whose sequence is known has a closely similar turnover number and, evidently, uses the same array of catalytic functional groups. Any amino acid change that is made by site-directed mutagenesis in any wild-type isomerase must, therefore, *either* have no effect on the catalytic properties of the enzyme, *or* lower the catalytic efficiency from its presently optimum value.

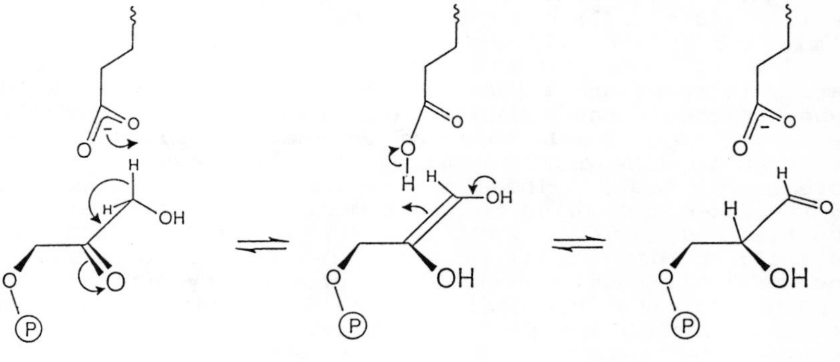

FIGURE 1. The reaction catalyzed by triosephosphate isomerase.

SITE DIRECTED MUTAGENESIS

In an early attempt to improve our understanding of this enzymic reaction, we changed the active site base from glutamate to aspartate, to produce the E165D mutant (3). This change, which moves the critical carboxylate by less than 1Å (as determined by Elias Lolis and Greg Petsko, comparing the crystal structures of the wild type and the E165D mutant chicken enzymes), lowers the k_{cat} value for glyceraldehyde phosphate by about 1000-fold. When the kinetic parameters for the mutant enzyme were deconvoluted, it became clear that the two enolization steps had each been slowed, largely because of a destabilization of the transition states for these steps. The proper interpretation of these large kinetic effects must, of course, await more detailed structural information on the E165D enzyme. Yet it is clear that, in this case at least, the relative movement of less than 1Å between a catalytic group and the substrate, is worth nearly 1000-fold in the rate (3).

In a second site-directed approach, we have examined the function of a mobile "loop" of ten residues of triosephosphate isomerase (from 168 to 177), that from crystallographic data appears to have a role in catalysis (4). In the unliganded enzyme the residues in this loop are poorly defined, yet when either a substrate or an inhibitor is bound to the enzyme, the loop is clearly fixed in a closed position. The closed loop shields the substrate from solvent and provides two main-chain glycine N-H bonds that bind peripheral oxygen atoms of the substrate's phospho group. Inspection of this loop suggested to us that four amino acid residues (from residue 170 to 173) could be deleted while producing only a minimal structural perturbation. Kinetic characterization of this deletion mutant has shown that the catalytic reaction is about 10^5-fold slower than the wild type. Although the substrate and product bind marginally less tightly to the mutant enzyme, phosphoglycolohydroxamate (the enediol analogue) is at least 10^3-fold more weakly bound. These data suggest that the function of the loop is to bind the *intermediate*, rather than the substrate or the product. It is as if the deletion mutant has lost its grip on the reaction intermediate, the enediol. This view is reinforced by our recent finding that the intermediate enediol is indeed lost from the deletion mutant rather readily. Whereas the wild-type enzyme "loses" less than one substrate molecule in 10^5, the mutant enzyme "loses" five molecules (that decompose very rapidly to P_i and methylglyoxal) for every molecule that is converted through to product (4). Deletion of these four residues has evidently turned triosephosphate isomerase into a methylglyoxal synthase!

The two examples cited above each provide new insights into the nature of enzyme-catalyzed processes, and illustrate the power of the site-directed approach in the investigation of enzyme mechanisms.

RANDOM MUTAGENESIS

In this second approach, we aim to learn more about the structure:function relationship by conducting an *unconstrained* search of the sequence 'space' of an enzyme, to see what mutations can produce an enzyme active site of high catalytic effectiveness (5). Our starting point for this work is the sluggish E165D mutant of triosephosphate isomerase described above. By subjecting the gene encoding this enzyme to random mutagenesis and then selecting bacterial transformants that carry isomerases of improved catalytic activity, we hoped to answer three questions: (*i*) can the catalytic activity of a sluggish enzyme be improved other than by simple reversion to the wild type; (*ii*) if such improvement is possible, is it rare; and (*iii*) how are such second site suppressor changes distributed?

There are many ways of introducing undirected changes into particular genes, ranging from the use of mutator strains or the induction of error-prone repair, to the use of a variety of chemical reagents that damage the DNA. Yet if our experiments are to be interpretable, we must be able to define how much of sequence space has been searched. That is, the random mutagenesis protocol that we adopt must be *definably* random, and the amino acid sequences generated must not be biased or limited by features such as hot spots or unreactive loops in the DNA substrate. Accordingly, we have opted for a technique that guarantees an unbiased search. Long oligonucleotide primers are synthesized that contain a fixed percentage of the three wrong bases at every position, and these 'spiked' primers are used in a standard mutagenesis protocol to produce a collection of mutants that span the length of the primer. By using several primers the whole of the structural gene of interest is subjected to mutagenic variation, the severity of which is determined by the percentage of 'wrong' amidite reagents used in the synthesis of the primer (6).

Ten overlapping oligonucleotides, ranging from 76 to 92 base pairs in length and spanning the entire gene for the E165D isomerase, were synthesized. At each position in the sequence (excepting only the aspartate codon at position 165), a 2.0 or 2.5% contamination of the three wrong amidites was introduced. Each of these oligonucleotides was then used to make at least 150 000 transformants in an isomerase-minus *E. coli* host, for each oligonucleotide window. These transformants were then subjected to a selection for those producing a higher isomerase activity.

From all of the generated mutants, which peppered the entire isomerase gene, six pseudorevertants were identified and characterized. The identity of the six pseudorevertants is indicated in Table 1. The catalytic potency of each of the purified proteins is improved over the starting mutant, by factors of between 1.3 and 19-fold.

As is evident from Table 1, the search for pseudorevertant isomerases from the E165D mutant has produced a variety of amino acid changes, scattered across the entire sequence, that can partially compensate for the primary lesion represented by E165D. The wide distribution of mutations suggests that the spiked primer approach is gratifyingly thorough in producing random changes in a target gene. What is also clear from Table 1 and Figure 2, is that the positions of all the effective changes are in or close to the active site. Thus, G10 lies three residues away from the active site lysine (K13), S96 and E97 are adjacent to the catalytic electrophilic histidine (H95), V167 is two residues from the essential catalytic base glutamate (which is aspartate in the sluggish mutant), and G233 provides a hydrogen bond to substrate (as deduced from the crystal structure of the yeast enzyme complexed with phosphoglycolohydroxamate). Of equal importance, *none* of those oligonucleotides that encode regions of the protein remote from the active site (e.g., numbers 2, 3, 5, 6, and 8) give rise to any pseudorevertants. While it is obviously premature to generalize, it may turn out for enzymes such as triosephosphate isomerase where there is no evidence for information transfer either within or between protein subunits, that only local changes in the first or second shell of amino acid residues near the active site will affect the specific catalytic activity. While the molecular details of these changes and the correlations with their functional consequences must await the appropriate structural results, it is clear that each of these local alterations nudges the active site back towards higher catalytic effectiveness.

TABLE 1.
SECOND SITE SUPPRESSOR MUTATIONS THAT IMPROVE THE CATALYTIC ACTIVITY OF A SLUGGISH MUTANT (E165D) OF CHICKEN TRIOSEPHOSPHATE ISOMERASE

Spiked oligo-nucleotide	Encoding amino acids:	Part of active site?	Isolated pseudo-revertants
1	1 - 26	++	G10S
2	25 - 52	-	none
3	51 - 77	+	none
4	76 - 100	++	S96P,or S96T,or E97D
5	99 - 126	+	none
6	125 - 151	-	none
7	151 - 179	++	V167D
8	178 - 205	-	none
9	204 - 228	++	none
10	226 - 248	++	G233R

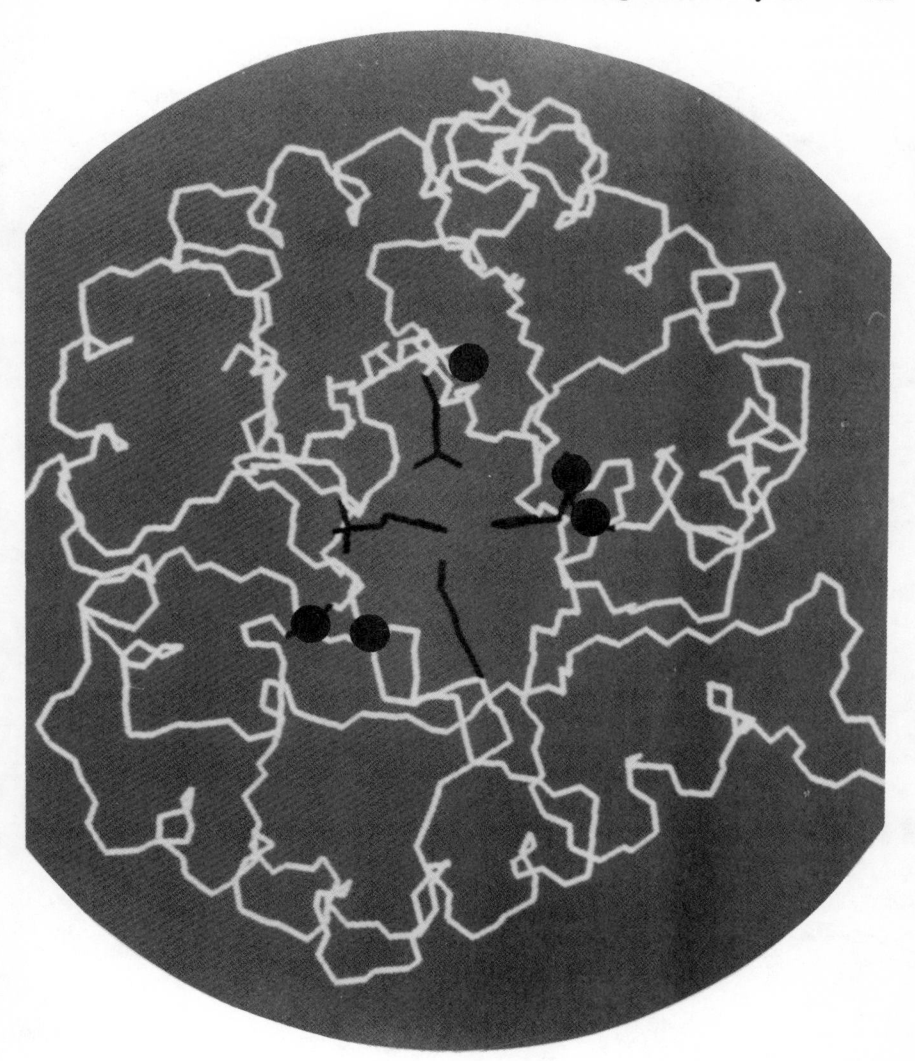

FIGURE 2. The structure of triosephosphate isomerase. The active site is in the center with the substrate in black. Above the substrate is the side-chain of Glu-165, to the right is that of His-95, and below is that of Lys-13. The positions of all second site suppressor mutations are highlighted with black dots.

REFERENCES

1. Knowles JR (1987), Tinkering with enzymes: What are we learning? Science 236:1252.

2. Knowles JR, Albery WJ (1977). Perfection in Enzyme Catalysis: The Energetics of Triosephosphate Isomerase. Acc Chem Res 10:105.

3. Raines RT, Sutton EL, Straus DR, Gilbert W, Knowles JR (1986). The Reaction Energetics of a Mutant Triosephosphate Isomerase in which the Active Site Glutamate has been Changed to Aspartate. Biochemistry 25 :7142.

4. Pompliano DL, Peyman A, Knowles JR. Stabilization of Reaction Intermediates as a Catalytic Device: Definition of the Functional Role of the Flexible Loop in Triosephosphate Isomerase. Biochemistry, submitted.

5. Hermes JD, Blacklow SC, Knowles JR (1987). The Development of Enzyme Catalytic Efficiency: An Experimental Approach. Cold Spring Harbor Symposia on Quantitative Biology 52:597.

6. Hermes JD, Blacklow SC, Koster H, Knowles JR (1989). A reliable method for random mutagenesis: The generation of mutant libraries using spiked oligodeoxyribonucleotide primers. Gene, in press.

Protein and Pharmaceutical Engineering, pages 167–177

EFFECT OF pH ON THE QUATERNARY STRUCTURE OF ASPARTATE TRANSCARBAMOYLASE[1]

David W. Markby, Edward Eisenstein[2], and H. K. Schachman

Departments of Molecular Biology and Biochemistry and Virus Laboratory, Wendell M. Stanley Hall, University of California, Berkeley, CA 94720

ABSTRACT A mutant form of aspartate transcarbamoylase (ATCase) was shown to have an average quaternary structure which was altered markedly over a pH range from 6.3 to 8.0. In the mutant enzyme, LYS 143 in the regulatory chain was replaced by ALA, thereby perturbing interchain interactions and destabilizing the compact, low-activity T conformation relative to the swollen, high-activity R form. The [T]/[R] ratio for r143ALA ATCase is 2.7 at pH 7.0 compared to 2×10^2 for wild-type enzyme. Changing the pH leads to alterations in the allosteric equilibrium for r143ALA ATCase resulting in [T]/[R] about 0.05 at pH 6.3 and a ratio greater than 70 at pH 8.0. Excellent agreement for these values was obtained from measurements based on the ligand-promoted changes in the sedimentation coefficient and from functional measurements of the cooperativity of binding a bisubstrate ligand. The regulation of ATCase activity by protons is interpretable directly by pH dependent changes in the $T \rightleftharpoons R$ equilibrium.

---

[1]This work was supported by NIGMS Research Grant GM 12159 and by NSF Research Grant DMB 85-02131. E.E. was supported by NRSA Grant GM 11067.

[2]Present address: Center for Advanced Research in Biotechnology, University of Maryland, 9600 Gudelsky Drive, Rockville, MD 20850.

INTRODUCTION

Although the dramatic effects of variations in pH on the activity of the regulatory enzyme, aspartate transcarbamoylase (ATCase, carbamoylphosphate: L-aspartate carbamoyltransferase, EC 2.1.3.2) from *Escherichia coli* have been known for more than 25 years, there has been relatively little effort directed toward correlating these functional changes with physical chemical studies of the allosteric transition at different pH values. As shown by Gerhart and Pardee (1), the wild-type enzyme is virtually devoid of cooperativity at pH 6.1 and the cooperativity is substantially greater at pH 8.6 than at pH 7.0. Moreover, the maximal activity, as measured by V_{max}, is increased almost 10-fold over this pH range. Analogous observations were made by Pastra-Landis et al. (2) who also showed that there was pronounced substrate inhibition at pH values greater than 7.8, thereby causing difficulty in obtaining reliable values of the Hill coefficient, n_H. In contrast to this unequivocal evidence for marked changes in both the cooperativity and maximal activity, there is little evidence about possible alterations in the quaternary structure of the enzyme which could account for the enzyme kinetics in terms of the well-characterized global conformational changes of ATCase as it is converted from the unliganded, low-activity, compact T form to the liganded, high-activity, swollen R state (3-10).

Such studies on wild-type ATCase are difficult to perform because at the lower pH values (about 6.0) the enzyme tends to aggregate in solution, and above pH 7.0 the T $\rightleftharpoons$ R equilibrium is so overwhelmingly toward the T state that accurate determinations of the [T]/[R] ratio cannot be made. These problems have been circumvented by using a mutant form of ATCase in which the T conformation is destabilized relative to the R form thereby permitting the detection of marked changes in the average quaternary structure of the enzyme over the pH range from 6.3 to 8.0. Moreover, through the use of equilibrium dialysis measurements (11) of the binding of the bisubstrate ligand, *N*-(phosphonacetyl)-L-aspartate (PALA) to the enzyme, we have been able to obtain quantitative data for the changes in cooperativity over this same pH range. In this way, the uncertainties inherent in studies of the enzyme kinetics have been avoided and the pH dependent changes in cooperativity have been related to alterations in the T $\rightleftharpoons$ R equilibrium.

SELECTION OF THE MUTANT

Although in principle ATCase mutants containing amino acid substitutions in the catalytic (c) chains of the enzyme could be used for these investigations, it seemed preferable to minimize the effects of replacements of residues near the active sites. Hence mutations were made in _pyrI_, the gene encoding the regulatory (r) chain of ATCase. In the unliganded enzyme three different interchain interactions are implicated in stabilizing the enzyme. One involves a principal contact between adjacent c and r chains, designated the c1r1 interface, a second between c chains in apposing catalytic (C) trimers, termed the c1c4 interface, and a third involving a c chain and an apposing r chain, designated the c1r4 interface. Since the crystallographic studies (6, 9) showed that the expansion of ATCase upon the binding of PALA, first detected by the decrease in the sedimentation coefficient of the enzyme (4), occurs at the expense of the c1c4 and c1r4 interactions, it seemed likely that amino acid substitutions at the c1r4 interface would have a significant effect on the T $\rightleftharpoons$ R equilibrium. In the R state the c1c4 and c1r4 interactions are virtually eliminated, and thus a weakening of those interactions in the T conformation through amino acid replacements would likely destabilize the T state of ATCase relative to the R conformation. Hence various amino acid replacements were made in the zinc-binding domain of the r chain (from about residue 100 to the C-terminus, residue 153).

As shown by Eisenstein et al. (12), the various mutants had properties markedly different from those of the wild-type enzyme. Several of the mutants were significantly less stable indicating the importance of interactions at the c1r4 interface for stabilizing the holoenzyme. Moreover, with some of the mutants it was possible to demonstrate that the rate of subunit exchange, produced by incubating the holoenzymes with free C subunits, was increased 10- to 20-fold when measurements were made in the presence of PALA. These results show that the PALA-liganded R state of ATCase is significantly less stable than the unliganded T conformation, in agreement with a conclusion inferred from measurements of the effect of PALA on the rate of disproportionation of an ATCase-like complex lacking one regulatory subunit (13, 14). The significance of interactions at the c1r4 interface was illustrated further by experiments on a mutant in which ASN at position 111 in the r chains was replaced by ALA. This mutant, designated r111ALA ATCase, exhibited neither homotropic nor heterotropic effects and was shown by sedimentation velocity experiments to be in the R conformation even in the absence of active site ligands known to promote the T $\rightarrow$ R transition of the wild-type enzyme (12). It appears, therefore, that the perturbation of interactions at the c1r4 interface can destabilize the T state significantly relative to the R conformation.

Although studies on r111ALA ATCase at different pH values might be useful, another mutant formed during the course of these investigations (Eisenstein, Markby and Schachman, in preparation) seemed more suitable because at pH 7.0 the enzyme seemed to consist of a mixture of molecules in both the T and R conformations. This mutant, in which r143LYS was replaced by ALA to give r143ALA ATCase, exhibited cooperativity in the binding of PALA but the extent of the homotropic effect, as measured by the Hill coefficient (n_H = 1.4), was significantly less than that of wild-type ATCase under the same conditions (n_H = 1.9). Moreover, the cooperativity was increased upon the addition of CTP (n_H = 1.9) and decreased in the presence of ATP (n_H = 1.0). From these PALA-binding isotherms it was estimated, based on the two-state model (15), that the [T]/[R] ratio was 2.6 in the absence of ligands, 37 when CTP was present and 0.14 in the presence of ATP. Moreover, direct determinations of the [T]/[R] ratio by difference sedimentation velocity experiments (4) yielded values of these ratios in excellent agreement with those evaluated from the equilibrium binding studies. Because of the apparent ease in altering the T $\rightleftharpoons$ R equilibrium of r143ALA ATCase by the heterotropic effectors, CTP and ATP, it seemed of interest to study this mutant enzyme at different pH values.

ALTERATIONS IN THE QUATERNARY STRUCTURE OF r143ALA ATCase AS A FUNCTION OF pH

The relative amounts of the mutant enzyme in the T and R states were determined by difference sedimentation velocity experiments. With wild-type ATCase, which is largely in the T state when unliganded (4), the binding of PALA promotes a 3.5% decrease in the sedimentation coefficient (16). This decrease is attributed to an increase in the frictional coefficient corresponding to a 10% increase in the hydrodynamic volume. In contrast, mutant forms of the enzyme which are devoid of cooperativity exhibit no change in sedimentation coefficient upon the binding of PALA to the six active sites (12; Newell and Schachman, unpublished). Mixtures containing equal amounts of wild-type ATCase in the T and R states have an average sedimentation coefficient which is 1.8% less than that of the unliganded, wild-type enzyme (17). Thus sedimentation velocity experiments yielding $\Delta s/s$ values caused by the addition of PALA provide reliable estimates for the fraction of protein molecules in the T and R conformations.

Figure 1 shows the values of $\Delta s/s$ as a function of PALA/ATCase for r143ALA ATCase at pH 7.0 (experimental data not shown), at pH 6.3, and at pH 8.0. At pH 7.0 the maximum value of $\Delta s/s$ is −2.55%, significantly lower than the value −3.5% observed for wild-type ATCase. When the pH is raised to 8.0 the value of $\Delta s/s$ caused by the binding of PALA to r143ALA ATCase is −3.5%, a value significantly different from that at pH 7.0. Conversely, $\Delta s/s = -0.25\%$ upon the binding of PALA to r143ALA ATCase at pH 6.3. From these values of the maximum $\Delta s/s$ we estimate that the [T]/[R] ratio for r143ALA ATCase is greater than 70 at pH 8.0 and decreases to about 0.05 at pH 6.3 as compared to 2.7 at pH 7.0.

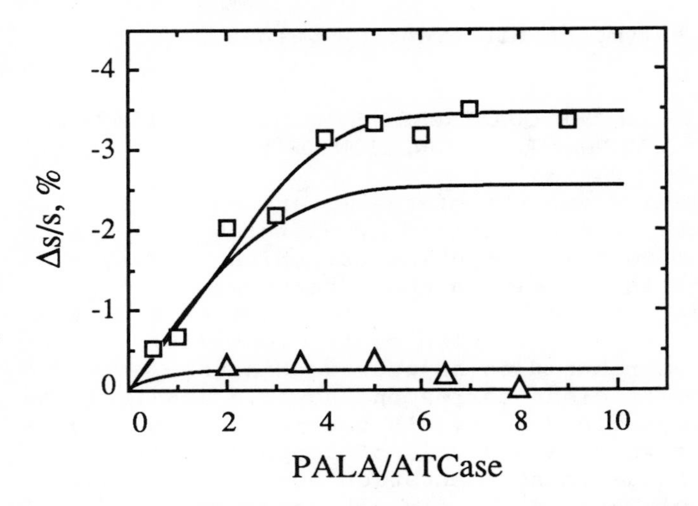

FIGURE 1. pH dependence of the PALA-promoted change in sedimentation coefficient for r143ALA ATCase. The change in sedimentation coefficient versus molar ratio of PALA/ATCase is given for r143ALA ATCase in phosphate buffer (40 mM potassium phosphate, 2 mM 2-mercaptoethanol and 50 μM zinc acetate) at pH 8.0 (□) and pH 6.3 (Δ) at approximately 3.5 and 3.0 mg/ml enzyme, respectively. The curve without data corresponds to pH 7.0 (from Eisenstein, Markby and Schachman, in preparation).

ALTERATIONS IN COOPERATIVITY OF PALA BINDING TO r143ALA ATCase AS A FUNCTION OF pH

In view of the structural evidence for changes in the allosteric equilibrium as a function of pH, it was of interest to determine how the affinity of r143ALA ATCase was affected by pH. Figure 2 shows the results of equilibrium dialysis experiments for the binding of PALA as measured by the method described by Newell et al. (11). The cooperativity at pH 8.0 is greater than at pH 7.0 with values of n_H equal to 2.0 and 1.4, respectively. In contrast, at pH 6.3 there is essentially no cooperativity in the binding of PALA as demonstrated by the virtually linear Scatchard plot and n_H of 1.0. Analysis of the binding data at pH 8.0, according to the two-state model, yields 790 for

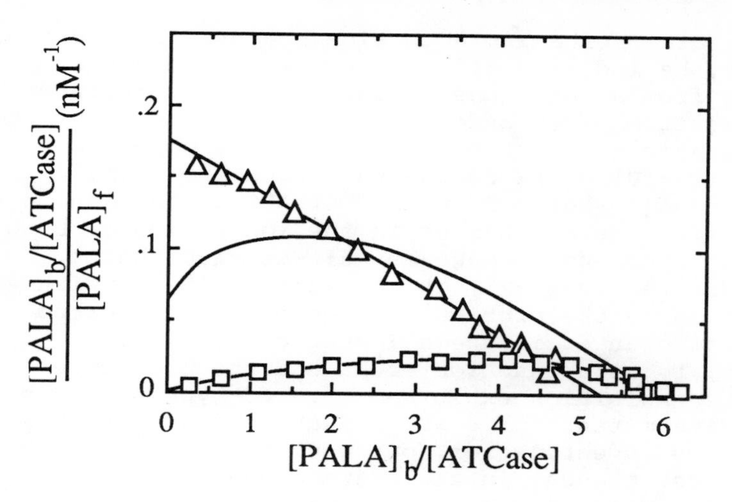

FIGURE 2. Scatchard plots for the pH dependence of PALA binding to r143ALA ATCase. Equilibrium dialysis was performed in phosphate buffer, either at pH 8.0 (□) using 0.4 μM enzyme or at pH 6.3 (Δ) using .05 μM enzyme. A reference curve (data not shown) was obtained at pH 7.0 Theoretical curves were obtained by fitting the model of Monod et al. (15) to the binding data as described elsewhere (Eisenstein, Markby and Schachman, in preparation).

the [T]/[R] ratio, K_R = 50 nM for the dissociation constant for PALA binding to the R conformation, and c = 0.025 where c is K_R/K_T and K_T is the dissociation constant for PALA binding to the T state. The data for the experiments at pH 6.3 yield a K_d of 29 nM and indicate that the number of binding sites decreased to 5.25 per enzyme molecule, presumably due to the poor solubility and aggregation of r143ALA ATCase at lower pH values.

DISCUSSION

Unravelling the effects of varying pH on enzyme catalyzed reactions is frequently difficult because of changes in the ionization state of the substrates, and alterations at the active sites because of changes in the ionization of amino acid side chains. In addition there may be indirect effects at the active sites stemming from alterations in the tertiary structure resulting from pH dependent variations in electrostatic interactions. Indeed, the observed effects (18-20) of pH on the variation in K_m and V_{max} for the nonallosteric C subunit of ATCase have shown the significance of changes in the ionization of both the substrates and analogs as well as side chains of amino acid residues (21) at the active sites. With the holoenzyme the problem is even more complicated because of changes in the allosteric equilibrium constant, L, as a function of pH and the pronounced substrate inhibition exhibited by wild-type ATCase and many of its mutant forms at a pH above 7.8. In this preliminary investigation, our goal was to relate the pH dependent changes in the average quaternary structure between the T and R conformations to the alterations in functional behavior of the enzyme in terms of its affinity for the bisubstrate ligand, PALA. For this limited objective, the mutant r143ALA ATCase proved ideally suited since, at pH 7.0, the enzyme contained appreciable amounts of molecules in both the T and R conformations (Eisenstein, Markby and Schachman, in preparation). Hence, shifts in the [T]/[R] ratio toward higher and lower values could be measured directly and reliably by the changes in average sedimentation coefficient resulting from PALA binding. In addition, quantitative isotherms for ^{3}H-PALA binding to the mutant enzyme at different pH values could be obtained by equilibrium dialysis (11), thereby avoiding the complexities and uncertainties inherent in interpreting the data from enzyme kinetics.

As seen in Figure 1, changes in pH have a marked effect on the maximal value of $\Delta s/s$ upon PALA binding. For r143ALA ATCase at pH 7.0, $\Delta s/s$ is −2.55% and this value increases to −3.5% at pH 8.0 and decreases to −0.25% at pH 6.3. By assuming that the maximal values of $\Delta s/s$ for the T and R states are −3.5% and 0%, respectively, we can make preliminary estimates for the [T]/[R] ratios. For r143ALA ATCase, L is greater than 70 at pH 8.0, 2.7 at pH 7.0 and about 0.05 at pH 6.3.

If the binding isotherms are interpreted in terms of the two-state model of Monod et al. (15), we calculate L = 2.7 at pH 7.0, L is about 790 at pH 8.0 and L is less than 0.05 at pH 6.3. For the latter pH, the binding curve is hyperbolic as seen by the linear Scatchard plot and, to a first approximation, L is essentially zero. These values of L, inferred from the binding data, are in excellent agreement with those calculated from the ligand-promoted changes in the sedimentation coefficient. It is of interest that the changes in pH had relatively modest effects on the affinity of the R state, K_R, for PALA; K_R is approximately 30 nM at both pH 7.0 and pH 6.3 and increases to only 50 nM at pH 8.0.

Clearly there is a need to perform binding studies on the isolated C subunit which thus far has been examined carefully only at pH 7.0 (11). In addition, it is important to note that all the experiments were performed in phosphate buffer and additional experiments are needed in an alternative buffer since phosphate is a product of the enzymatically catalyzed reaction and is known to bind at the active sites. How the competition between phosphate and PALA binding varies with pH is not known. In addition, the relative affinity of the T and R states of ATCase for phosphate in different ionization states and ionic strengths has not been investigated. Despite these uncertainties the results, obtained from both structural and thermodynamic studies, demonstrate that lowering the pH from 7.0 to 6.3 stabilizes preferentially the R conformation of r143ALA ATCase. Conversely, raising the pH to 8.0 leads to a preferential stabilization of the T conformation. The excellent agreement between the sedimentation velocity and binding experiments provides powerful evidence for the regulation of ATCase activity by protons, as well as nucleotides, through preferential stabilization of either the T or R state. Moreover, these conclusions account for the changes in cooperativity observed in enzyme kinetics of the wild-type enzyme as a function of pH (1, 2). These observations have a bearing on the interpretation of crystallographic studies on ATCase which thus far have been conducted only on the enzyme at a pH about 6.0.

REFERENCES

1. Gerhart JC, Pardee AB (1963). The effect of the feedback inhibitor, CTP, on subunit interactions in aspartate transcarbamylase. Cold Spring Harbor Symp Quant Biol 28:491.
2. Pastra-Landis SC, Evans DR, Lipscomb WN (1978). The effect of pH on the cooperative behavior of aspartate transcarbamylase from Escherichia coli. J Biol Chem 253:4624.
3. Gerhart JC, Schachman HK (1968). Allosteric interactions in aspartate transcarbamylase. II. Evidence for different conformational states of the protein in the presence and absence of specific ligands. Biochemistry 7:538.
4. Howlett GJ, Blackburn MN, Compton JG, Schachman HK (1977). Allosteric regulation of aspartate transcarbamoylase. Analysis of the structural and functional behavior in terms of a two-state model. Biochemistry 16:5091.
5. Krause, KL, Volz, KW, Lipscomb WN (1987). 2.5 Å structure of aspartate carbamoyltransferase complexed with the bisubstrate analog N-(phosphonacetyl)-L-aspartate. J Mol Biol 193:527.
6. Kim KH, Pan Z, Honzatko RB, Ke HM, Lipscomb WN (1987). Structural asymmetry in the CTP-liganded form of aspartate carbamoyltransferase from Escherichia coli. J Mol Biol 196:853.
7. Kantrowitz ER, Lipscomb WN (1988). Escherichia coli aspartate transcarbamylase: The relation between structure and function. Science 241:669.
8. Schachman, HK (1988). Can a simple model account for the allosteric transition of aspartate transcarbamoylase? J Biol Chem 263:18583.
9. Ke H, Lipscomb WN, Cho Y, Honzatko RB (1988). Complex of N-phosphonacetyl-L-aspartate with aspartate carbamoyltransferase. X-ray refinement, analysis of conformational changes and catalytic and allosteric mechanisms. J Mol Biol 204:724.
10. Allewell NM (1989). Escherichia coli aspartate transcarbamoylase: Structure, energetics, and catalytic and regulatory mechanisms. Annu Rev Biochem Biophys Chem 18:71.
11. Newell JO, Markby DW, Schachman HK (1989). Cooperative binding of the bisubstrate analog N-(phosphonacetyl)-L-aspartate to aspartate

transcarbamoylase and the heterotropic effects of ATP and CTP. J Biol Chem 264:2476.

12. Eisenstein E, Markby DW, Schachman HK (1989). Changes in stability and allosteric properties of aspartate transcarbamoylase resulting from amino acid substitutions in the zinc-binding domain of the regulatory chains. Proc Natl Acad Sci USA 86:3094.
13. Subramani S, Bothwell MA, Gibbons I, Yang YR, Schachman HK (1977). Ligand-promoted weakening of intersubunit bonding domains in aspartate transcarbamoylase. Proc Natl Acad Sci USA 74:3777.
14. Subramani S, Schachman HK (1980). Mechanism of disproportionation of aspartate transcarbamoylase molecules lacking one regulatory subunit. J Biol Chem 255:8136.
15. Monod J, Wyman J, Changeux JP (1965). On the nature of allosteric transitions: A plausible model. J Mol Biol 12:88.
16. Howlett GJ, Schachman HK (1977). Allosteric regulation of aspartate transcarbamoylase. Changes in the sedimentation coefficient promoted by the bisubstrate analogue N-(phosphonacetyl)-L-aspartate. Biochemistry 16:5077.
17. Werner WE, Schachman HK (1989). Analysis of the ligand-promoted global conformational change in aspartate transcarbamoylase: Evidence for a two-state transition from boundary spreading in sedimentation velocity experiments. J Mol Biol 206:221.
18. Weitzman PD, Wilson IB (1966). Studies on aspartate transcarbamylase and its allosteric interaction. J Biol Chem 241:5481.
19. Allewell NM, Hofmann GE, Zaug A, Lennick M (1979). Bohr effect in Escherichia coli aspartate transcarbamylase. Linkages between substrate binding, proton binding, and conformational transitions. Biochemistry 18:3008.
20. Leger D, Hervé G (1988). Allostery and pK_a changes in aspartate transcarbamoylase from Escherichia coli: Analysis of the pH dependence in the isolated catalytic subunits. Biochemistry 27:4293.
21. Kleanthous C, Wemmer DE, Schachman HK (1988). The role of an active site histidine in the catalytic mechanism of aspartate transcarbamoylase. J Biol Chem 263:13062.

Index